The mapmaker

Maitry Roy Moulik

Translated from Bangla by
CHANDANA DUTTA

THORNBIRD
NIYOGI
BOOKS

Published by
NIYOGI BOOKS
Block D, Building No. 77,
Okhla Industrial Area, Phase-I,
New Delhi-110 020, INDIA
Tel: 011-26818960, +91 7428978881
Email: niyogibooks@gmail.com
Website: www.niyogibooksindia.com

Editor: Anwesha Panda
Design: Nadeem Ahmed
Cover design: Antra K.

ISBN: 978-81-992494-9-3
Publication: 2025

This translation in English is published by arrangement with Ananda Publishers Pvt. Ltd. who originally published in Bengali in 2022.

Printed at Niyogi Offset Pvt. Ltd., New Delhi, India

Ramapithecus and I

The young swamp he came to
Six million years ago,
His unfazed mother beside him,
His father recently dead,

Is the wall map's mixed forest,
A dotted power line along its edge,
And the window's low, clouded hills,
Dolomite and fossil rich.

Making his home
Where his implements took him,
He waited for the rains to break.
Cutting my finger, it's his blood I taste.

—Arvind Krishna Mehrotra

Arvind Krishna Mehrotra, Collected Poems: 1969–2014, Penguin Random House, India, 2014. Reprinted by permission of the author.

CONTENTS

THE BEGINNING

The full moon stains itself
Yet who dares call it spotted.

"Disgrace", Rabindranath Tagore

It is the first half of the nineteenth century. A cartographer has gone missing. He was last seen outside his own house in London's Buckingham Street. Broke and penniless, just released from prison, he was trying to enter the five-storeyed building where he lived. Hired sentries had held him back. The man was no ordinary cartographer. Geographical maps had long been available with people. But this man was different, he had made geological two-dimensional maps marked with each layer of rock below the ground for the whole of England, South Wales and Scotland.

For the enlightened classes of England, the concepts of the Renaissance and the subject of science had just about become synonymous. The Archbishop of Armagh, James Ussher, had determined the six-day creation-of-the-universe theory which was now under severe and continued criticism by Erasmus Darwin, grandfather to Charles Darwin, at the monthly meetings of the Lunar Circle Club held in the presence of James Watt and several other intellectuals. There seemed to be no evident scientific logic to the bishop's claim. In which case, how long had the universe taken to be born—six days, six years, six hundred years, or perhaps centuries?

But irony stares in the face of unravelling truth. The man who has toiled tirelessly to unearth and present evidence of how the earth's face has altered over millennia cannot, rather will not, be acknowledged for his contribution because he has no pedigree to boast of. He is unschooled, destitute and rustic, the son of a blacksmith, and not, therefore, worthy of sharing a table with a duke, earl, knight, or a baron. And yet, many others are singing paeans to him.

The phase of time brought to light in this novel are the years before and after 1819—when the geologist William Smith went missing from London. By this time the England-Scotland geological map had already been published by the Geological Society of London. The cartographer acknowledged as the creator of this map was George Bellas Greenough, the President of the Society. Many considered this map fake. They believed the real mapmaker to be William Smith, the son of

John Blacksmith of village Churchill. But where was William Smith? Why had he vanished without a word of protest or explanation about this gross injustice and theft? And where were his mentally deranged wife and orphaned nephew? Would William ever return? Would aristocratic bluebloods like Greenough ever let him return?

This is the story of a mapmaker, the mapmaker of the first map that would change man's perspective about the creation of the earth. It will take us on a journey through a sliver of time between the end of the eighteenth century and the early nineteenth, and help us travel through the dark dappled history of a country we know as England.

YEAR 1819, LONDON

It has poured relentlessly for about six days now, a weather some enjoy, and some do not. The rain has brought with it a dream-like drowsiness. The city gleams like newly washed glass. It is spotted wet, here and there, and yet shines pristine in its brightness. Inside, the room is strong with a damp smell, that crawls out through the gaps to tug at the feet of passers-by. In the last twenty years, the city has swollen twice over in population. It has witnessed a stream of people flow into it from the countryside, transforming in these passing years into a supposedly gigantic wish-fulfilling desire-granting tree. But perhaps it is not this, not as the stories would have us believe. It seems that if you hold out your hands in the city air, pounds and shillings will come cascading down, filling your cupped palms. No one goes hungry here. Business and money come your way even as you walk about. But is this even possible?

A smirk escapes the man as he ponders. There's the bridge over the Thames. He directs his feet against the flow of people.

Then he begins walking northwards. The rain has not stopped yet. Many still hold umbrellas over their heads, especially the women, their fashionable, brightly coloured ones, trimmed prettily with lace. Half-hidden faces peep from under them. Can you call a woman beautiful just by looking at her chin? Perhaps, yes. As for the girl who went by just now, the one with her mauve flower-sprinkled umbrella with its broad woollen lace frill, he surmises, taking a quick look at her feet and waist, that she is not yet thirty. Her lips are a rich pomegranate red. Her lashes are perhaps lush and long, as they throw a hint of a shadow on her nose. But maybe not, maybe it's all just his fancy, and yet his eyes have never betrayed him. It's been forty years, and he has lived his entire life with utmost faith in what his eyes see. The same eyes, that many years back, had halted upon a poundstone at his uncle's dairy. And since then, what they have witnessed, no one else has.

No, no, it is all a mistake. If not, would he be out wandering these streets as a destitute!

He pauses at the head of the bridge trying to look back. The tower of King's Bench Prison can no longer be seen. He hasn't met anyone since his release from there. No one had come to receive him. At the beginning of his prison sentence, his wife would visit. And sometimes she would carry his nephew along. His insides would be in a whirl whenever he saw her, 'Did you meet Cary? Did he say anything about getting me out? Didn't he ask how much it would cost?'

His wife would look on, dazed. She had nothing to say. That helpless look would make him furious, set off his griping. 'You are utterly useless. Such an old friend of mine and you can't even find time to tell him of my plight!'

She would lower her gaze, 'He's…not there.'

'Not there? Must be busy. He doesn't know I'm here. Or else he would have paid off all my debts and taken me back.'

'Oh! Okay.'

'Now what does that mean? Don't you realize how important your husband's work is? The history of the earth will be overturned. No one will believe any longer in the tales told by the church. If only I could get out of this jail. Go, look for him, no matter what, find Cary. Don't stand idling. You're not planning to run off with someone, are you?'

The wife would remain silent. A while later she would say, 'I'll take your leave.'

A day at a time went by, and she came by every day, for a whole week. On the last day, she brought his nephew too. She stretched her hand to him, holding a few slices of cake wrapped in a torn napkin. He'd pitched it back, almost hitting her face. 'I'm rotting here in this cursed prison, and you come to feed me cake. Does it not occur to you to bring some money? Why can you not do anything well? Be off now, try and earn something. Get out of my sight. I cannot bear you any longer.'

The two of them left and never came back. The first day after this, he found their absence unbearable. It made him desperate to pull out his hair. If only he could howl his lungs out, 'The wench has bolted. Women are like pigeons. When the home lacks food, the pigeon flies off. It's utter rubbish how she's not right in the head. She feeds me cake while dallying with others. Why should she live with this beggar? I have feasted at my mother's wedding in childhood, and now I'll feast at my wife's.' But she never came back, never again. He paced wildly behind the bars. Then he thought, I'll bash her head in if I see her again. Then again, he thought he had been wrong to have yelled at her. He'll beg her to forgive him. Thoughts whirled around in his head. After a while, he stopped thinking. He went blank. Like so many others here, his arms splayed out of the iron bars, he began to wail, 'Sir, I beg you, show some mercy. I am a gentleman. I couldn't pay off my debts and they threw me in. I haven't killed anyone, nor have I stolen anything. Pray, sir, if you don't show mercy, I will rot here. I am…I…the future…'

He has now composed himself. No, he is nothing to the future. What future can he be, he who has lost the right to live with dignity? Is he the son of a duke, or the grandson of a baron? He is not even the heir to an industrialist! No one cares if he lives or dies.

There is nothing more for him to seek now. And yet he walks on. He himself doesn't know why. The gaslights along the bridge are still glowing. As long as the skies remain overcast, the lights will burn on even in the morning. At the end of the bridge, the man takes the road to the city. He walks on up to house no. 181. He stops, frozen. This is Cary's office. How often he would visit it in the days gone by. He had come here even the day before he was thrown into prison. Cary is a valuable ally, a man who understands the trade, affluent and established. And the best of all, Cary is never swept away by his emotions unlike him. But he cannot linger here any longer.

He strides on. He has just turned fifty, how can he let himself grow decrepit and diseased? It hasn't even been some months since he dashed across the ends of this country. He has toiled through nights, with no inkling of hunger or sleep or anything whatsoever.

He walks past Savoy Street and soon past John Adam Street. At last, he stops in front of a house on Buckingham Street. House no. 15. He can never forget this address; he has lived here for fourteen long years. He pushes the front door. It is tightly shut. But not from the inside. He now notices the large lock on the latch outside. Oh! The landlord must have taken over the house again. Have they taken away his belongings too? How were his debts repaid? Did they sell off his valuables, his furniture, and all that he possessed? Spotting a notice on a wooden plank, he walks up to plod through the information, reading one word at a time. His eyes have become unused to light, they've been in the dark too long. He wouldn't otherwise take so long to read a two-line note.

Outsiders are not allowed inside the premises. Forced entry will be punished by law.

What manner of notice is this? How is his entry forced? He was living in this very house for so many years, his staff, too. Nobody had ever entered it forcibly then. Who is this notice for?

A guard comes sprinting towards him. Looking him up and down, he questions, his manner rough, 'May I have your name, sir?'

He answers, 'William. William Smith.'

The guard snickers, his irritation bordering on contempt. 'You cannot go in, sir. This notice is especially meant for you. Please do not attempt to come in at all. I will be compelled to call the police.'

The man stares at the guard blankly. Then he turns around and heads back the way he has come. But where should he go? He must first locate his wife and nephew. He could stop by at a few known addresses. He might find them at one. But then, he might not. Which woman will wait this long for her man to turn up? Certainly not the one whose husband has plagued her with woes her entire life. He has given her nothing, no home or hearth, no fineries to adorn herself, not even a child to call her own. All her life she has run around at his beck and call. What has she not endured for his sake, from surveillance by the police, to harassment by servants of the royal household, as also taunts thrown their way by the court and its jury.

A woman who still awaits her beast of a husband after all this, is indeed noble. But then she is not noble, his wife, she is just an ordinary woman. And mentally ill. How long has it been since he looked upon that pale, pallid, weary face? Never, he will perhaps never find her. She is lost to him. But what if he does find her? Although, in his own life, he could never change anything impossible into the possible. If he can indeed make this one thing happen, if he does come upon her, he will then not stay a minute longer in this abhorrent city. It has given him nothing but shame and neglect, day after day. He will leave for his village home, a tiny village that has raised itself around a little church. It is time to leave London. He will leave, even if alone, but not without seeking out his wife and nephew at least one more time. He will make that last attempt to find them.

It might do good to visit Cary's office once. He may find a clue there. But no, he will never go there again. How will it help in any case? Why would a wealthy publisher remain friends with a blacksmith's son? It worked while there was work, but now there is none. If indeed he had been a true friend, he would have visited King's Bench. What his wife could not express then, he can fathom fully now. He must absolutely find her. It could not all have been for naught. There is always a last resort. Dear God, is a last resort waiting for me? He laughs as soon as he thinks this. People will wonder at his madman's laugh. He looks up at the sky and mutters, 'Dear God! Were we ever rivals, do you think? I never took a bet to prove you false through my work, I only wanted to reveal the truth. Do you not wish man to know the truth? If I find Mary and John this day, I will know that you wish it. If not, I know not what the future holds. Goodbye Cary, for walking along with this baseborn, thank you indeed. You were never at fault, my dear friend. One cannot stand stubbornly against the entire upper class of our society for long. But we have lost all, the record of our work done through so many years, our papers, dairies, notebooks. And they have finally managed to show me for who I am, a common man. I am destined to fail, I can never cross the line beyond food, sleep and back-breaking toil.'

And that day, the man made a serious mistake. If only he had stopped for a moment at Cary's, before he left London, he would have witnessed a future history in the making, a history to which he had laid the foundation.

A map! Eight and a half feet in length, six and a half in width. A thoroughly drawn, intricate canvas marked with landforms and substrata of England and South Wales. The earth on which men tread so thoughtlessly holds in its bosom a hoard of treasure. Rocks and stones that are not a day or a week old, but ancients birthed millions of years ago. Their true beauty blossoms through these maps. It is Wednesday tomorrow, the first day of September. The day when the world's very first geological atlas, now in the form of a book, will make its journey from Cary's office. It's been cartographed by John Cary, assisted by his son George. Four years ago, when the first print of these maps was published, they didn't sell beyond a few copies. Tomorrow the markets will receive an atlas with a detailed layout of the nation's geological contours, its main map accompanied by a few fresh additions. The true credit for all of these is owed to William 'Strata' Smith.

As to Smith's whereabouts, Cary is clueless. As he is to what the morrow will bring. Will he be rewarded with accolades or thrown into prison to be beaten to death? Will a gang of radicals bring down his house with stones? After all, the faith of years is about to crumble away in bits and pieces. That is no mean matter. There will most certainly be retaliation. This fear has kept him away from Smith. He knows of Smith's release from prison. He also knows how Smith has been forbidden from entering his own house. What he doesn't know is where Smith is now.

***

At dusk the next day, a coach drawn by a pair of horses pulls quietly out of London. The window shutters are tightly sealed from within. The coachman has been instructed to stop only at the Black Swan Inn in Holborn. He doesn't know where he will head after that. Those who have hired the carriage are probably not from a well-to-do family. There's an elderly gent, a woman, and a young lad. Not a word passes between them. The woman has her head placed on the old man's shoulders, her eyes glazed and helpless. Perhaps she is his wife. She was heard whimpering for a while. They are not vagabonds, and yet they could be seeking refuge somewhere. Will they be able to pay the coachman off? He must recover a part of his money as soon as gets to the Black Swan. And then, he will insist on being paid for every bit of the journey covered. Do they even know where they are headed? The coachman sighs deeply. The city of London stands a silent bystander as the numbers of such impoverished unfortunates swell, day by day by day.

YEAR 1769, CHURCHILL VILLAGE

A couple, cradling their newborn, stands in front of the church in their village. It's a tiny village called Churchill in Oxfordshire. The villagers begin their day with the first toll of the carillon. Today is special. John and Anne, native to Churchill, have brought their firstborn to church for his baptism. John Blacksmith is reputed throughout Oxfordshire as a skilled worker. Whatever he forges—cups, spoons, bowls, spades, sickles—lasts long. So sharp are his knives and cleavers that one is hard put to find similar ones even in London. Though to tell the truth, most Churchill residents have never been to London. But they have used wares from other smiths in the neighbourhood, and these are nothing like John's. The way he fits bolts to metal, tools stand their shape even when thrown down the mountainside. And now John's child will be given a name after being baptized just as all the other village children have been named by their church.

John's brother lives in the next village. He will join them for the ceremony today. The brother is doing better than John. He even owns a few acres of land and has dreams of starting a dairy. It seems fine-breed cows are being sold these days. And all sorts of new and improved farming equipment are being produced, too. Of course, none of this can be seen in the Churchill market because it is not enough to simply procure these. Farmers will not buy them unless they find the prices suitable. And these can't be produced in small smitheries. They require large places to be made in, places with fancy names called factories. Such factories employ huge numbers of workers. If only he could put some sense into his brother and get him to join a factory rather than spend his days in this tiny village sweating at the furnace. A talented lad, John would be a catch in the city. But he's failed to drum any sense into him so far. If only he could get him to change his mind now.

Slowly, people are gathering around. The priest hasn't arrived yet. As the conversations grow in little groups, trouble seems to be brewing. Not trouble exactly, but an argument of sorts. Several potters live to the east of the village. One of them is Richard, Richie to the village folks, who molds beautiful glass plates. Richie has spent the last six months in a Hanley factory, learning to better his craft. The quarrel, which had begun about the advantage of factories, has now taken a different turn.

'What is this cursed place, this factory! The Lord God's given us hands and legs to work with, not to vegetate. It's the devil's work to create machines and suchlike,' Peter, the village tailor snaps.

'You can't dismiss it like that, you old man. You'll know what a factory is only when you see it with your own eyes. It's teeming with people, working away, row after row, building the same thing. How many plates can a potter make in a day? At most twenty. And then he needs to dry them, paint them, and dry them again. Go, have a look at this machine that makes two thousand plates in two days. Each one an exact copy of the other, nothing knobby or uneven about it.'

'What folly is this! Has God made us all in the same fashion? No, he hasn't! Why should a machine do it then? This is a sin that cannot be forgiven, brothers.'

'What sin are you talking about? City folks are happily eating off these plates. They seem to be doing fine.'

'Wait and watch. God will punish them for their sins. Their bodies will break out with sores, and they'll be struck down with leprosy.'

'Oh, let it go, Peter. This one's a city boy now, he's lost his head don't you see? How do we profit from arguing with him?' A bearded man, about Peter's age, walks by as he says this.

'Look, brother, I'll not waste my time over what happens in the city. But the trouble is that our boys are coming back with their heads ruined. These days it's all about being modern. Isn't it enough that we've seen automobiles with wheels? Let the factories be.' Peter's voice continues to rise.

'Lower your voices, all of you. Father's coming. Don't speak of this in his presence,' someone pipes up.

Everyone falls silent.

The priest arrives. He picks up the newborn and pours water on his head gently. Soon, he completes the ceremony and hands the little one back to John and Anne. And then he makes an announcement. 'Henceforth your son will be called William, born 23rd of March of the year 1769. Exactly five thousand seven hundred seventy-two years four months and sixteen days after God created this world. Put this in bold letters outside your house, so no one forgets how long it has been since God created this world and William was born. Have you paid heed to what I've said?'

'Aye, Father.'

'On this joyful day, it is important to state further and denounce the unbelievers spreading untruth about our Lord. They do not wish to abide by the beliefs of the church. They reject the faultless prediction of our great Bishop James Ussher. So be it. I do not wish to mar the purity of this happy occasion today by speaking of sinners. But I intend to remind you again of the valuable words of our Reverend, the Bishop, which all of his fraternity consider infallible.

My dear residents of Churchill, do not forget that four thousand four years before the birth of Christ, at the break of dawn on a Monday, as the clock struck nine, this earth was created. Speak in one voice and answer, who was its creator?'

'The Lord God.'

'Very well said. In the beginning, God created heaven and then the earth. And the earth was without form, and void, and darkness was upon the face of the deep. And the Spirit of God moved upon the face of the waters. Tell me now, in which month was the earth created?'

'October.'

'Aye. In the last week of the month of October, this earth was created. It was dark at first, and then there was light. And then the sun and the moon. And came the oceans and seas, the rivers, green valleys, deserts, mountains, snow peaks, every little thing that lives on the earth. And then the strangest of wonders took shape. The earth came alive with the breath of life. Speak out, what date was it?'

'26 October.'

'Four thousand years before the birth of Christ, on 26 October. On this holy morning, God beheld how the earth had come alive with the breath of life. By the evening many creatures took life—spiders, scorpions, eagles, cheetahs, horses, monkeys. These walked, ran, jumped, crawled, slithered. Can you see it in your imagination? God looked upon them with a smile as all his creatures wandered about. Dear God, bless these simple folk with the gift of divine sight that they may all see your wonder. My fellowmen, do you behold as I do? The vision of that day. All will see save the sinners.'

'Yes, Father, we do!'

'Amen. How wondrous is your play, dear Lord. The very next day plants bloomed, then the meadows. And roses, apples, palms, pines, daisies. Heaven descended on this earth. In this second heaven, the Earth, the Lord allowed even those who disobey him. Can you name the first two?'

'Adam and Eve.'

'But Adam and Eve were not human, they were heavenly beings still. They were denied heaven because they had sinned. That Saturday midnight, the Lord placed these two in the midst of his creation, and thus he finished his work. How many days had passed by now?'

'Six.'

'Hear this you virtuous souls, God created this earth in six days. Only six. Do not doubt this ever again. Should you doubt, you will be a sinner too, who doubts the very existence of the Lord, an infidel, who deserves hell and will burn in its furnace of fire. You will scream and shed copious tears but will receive not a drop to drink. Terrible torment will visit you. But sinners deserve to be punished.

So leave all doubts aside, do not disbelieve. Is there anything you wish to ask?'

'No, Father.'

'Gather together then and let us sing to the glory of the Lord.'

The congregation is deep in song, hands folded, eyes closed. But there is someone who cannot be at peace, as his mind wanders about restlessly. It is John Blacksmith's farmer brother. He is sitting at one end of the long church bench, while Richie sits at the other.

What if Richie slips out the side door as soon as the prayers end? Why would he wait around after the way he was hounded earlier by some of the village folk? He might even leave the village forever now.

And indeed, Richie is walking away. The man now tries to hurry past the crowd as best as he can. But it's not easy. Running along breathlessly, he at last catches up with Richie near the swamp. Fortunately, there's no one around.

'Listen, my good fellow.'

Richie turns around. 'Whatever are you running for?'

'To catch up with you.'

'Me? Do I look like a thief or a robber? Ah, now I know. You didn't like what I said this morning? Forget it, I should have just stayed quiet.'

'No, not at all. I wanted to ask about Hanley pottery. A friend left two bowls with me, fine and smooth. Our potters can't make those. The bowls are from the factory there. What are they called?'

Richie stares surprised. John's brother is still panting. 'Very fine bowls? You can see the food through? But doesn't crack under the heat?'

'Yes, exactly. He'd mentioned a name, am forgetting it.'

'Fine pottery.'

'Yeah! You work there, don't you? What's your owner called? What a great enterprise he has set up.'

'Josiah Wedgwood.'

'I remember now. I've read of him, in the papers.'

'I say, can you read?'

'A little. This bloke, the friend who brought the bowls, carries newspapers sometimes when coming down from London. I'd read about Wedgwood in it. Extraordinary potter, it said. He's changing pottery completely with his factories. He's bowled London over.'

'Why just London, the whole world's now hearing of him. Do you want to see his factory? I can get you work there. He needs a lot of workmen.'

'Nah, I'm no potter, you're looking at a farmer. I've heard they are also making new equipment for the likes of us, machines that can till land in a day. One can sow seeds quicker.'

'That is true. Have you heard of John Wilkinson?'

'Wilkinson? Who's that? A potter?'

'Nah,' Richie interjected, 'not a potter, but an ironmaster. Abraham Darby and John Wilkinson are industrial iron casters. Built bridges don't you know?'

'Iron bridges, really! How can they cast iron so thin for it? Now you're telling me stories.'

'Nah. Come with me, I'll show you. They've built an iron statue and a chapel, completely with iron.'

'You've seen all this!' John's brother was suspicious.

'Where do I have time for that? But I've seen the iron loaders carrying molten iron across. The loader wheels are iron too.'

'Even the spokes?'

'One can't talk to you if you are so distrustful. Better you go and join the village festivities.'

'You misunderstand me, it's not mistrust. I can see the country's changing over. It's becoming new, they say modern. I too know a little. Some James somebody is creating steam now.'

'Aargh! What nonsense. Not steam. That you see in your house too, don't ya? It comes out of your mouth in winters. But an engine that runs on steam. That! It can pull anything that is hooked up to it. You won't need a horse, or bull or camel anymore for that.'

John's brother looks stumped. 'Dear God! I can't think anymore. How are these people creating such wonders?'

'Why is it so difficult to see? All of you closed your eyes to see what God created, the horses, monkeys, men, as the Father said. So what is the problem now? You can't think now, but very soon it'll all be in front of your eyes. I say, you didn't hesitate to believe what God managed in six days without seeing it for yourself, so why do you balk now?'

'This is where you go wrong. You disbelieve in God's powers. That's why people condemn you. No, no brother, do not compare God to anyone. That's what sinners do. You're a father to children, how can you keep repeating this sin?'

'Forget it. I won't say it again. Forgive me.'

'I can't, you'll have to ask for His forgiveness. I too have sinned by listening to you. No, no, let us ask for forgiveness tonight before we go to bed. Will you do it, tell me?'

'I have sinned, so I will.'

'That's a relief. Now tell me, will we ever see the farming equipment in these parts? Will the King never look upon us in the villages?'

'That's in the future. How can we say what will happen then? If you are too restless, go once to the city. You might understand. They've also built a spinning frame, that runs with water power.'

'With water? The water will move the wheels?'

'That's what I heard. Someone called Richard Arkwright has built it. The frame will rotate as the water falls, and the machine will spin yarn. I haven't seen it yet, just so that you don't ask. But I think it to be true. After seeing my master's factory, I can believe in the unbelievable.'

Richie walks away. John's brother can't leave the village as easily to visit the city, his land is waiting to be tilled. If only he could get John to go instead. Such an obstinate fool, not excited by anything new. It's his fate to stew by the fire, hammering away. But his own future doesn't seem promising either. If the winds of change do not blow through these villages, they'll all be condemned to remain as they are. He cannot even think of leaving the village just for work.

He turns around, crestfallen, and drifts back towards the church.

YEAR 1780, BIRMINGHAM

Birmingham, a picture-perfect city, north-west of London. A hundred and twenty-six, maybe twenty-seven, miles from it. A city of the elites, intellectuals and the uniquely talented. It publishes its own newspaper every Thursday, *The Birmingham Gazette*, which everyone, right from the Royal House of Commons to dukes and barons, looks forward to.

There's a three-storeyed white mansion in the city, Soho House. Tonight, Soho House will witness a conclave of some of the country's most brilliant minds. Indeed, it is the night for the monthly Monday meeting for members of the Lunar Circle Club.[1] Soho House has almost always hosted these meetings, barring once or twice.

1 The Birmingham Lunar Society was a dinner club and an informal learned society of prominent figures from England, including scientists, industrialists, natural philosophers, and intellectuals, who regularly met every month. The Society held forty-eight such meetings without a break from 1765 to 1813.

This august gathering will continue in this venue as long as its owner, Matthew Boulton, is alive. Boulton Estate covers a wide expanse of land, carpeted beautifully with green grass and enclosed on all sides with apple and cedar trees. And in its centre, the stunning mansion, glowing bright and white like a majestic swan. It's a full moon tomorrow. Lunar Circle meetings are held very close to full moon nights, maybe a day before or after, sometimes on a Sunday. The reason? The moon's light makes the journey of the members easier in the absence of streetlights. The members jokingly call themselves 'Lunaticks', those who think up crazy ideas which the whole of England pays for. It has fourteen such loonies right now. Just as ordinary folk cannot be its members, so can't dukes or barons. Even though members of the club are highborn aristocrats, they are inimitably different from other title holders.

The lord of the estate, fifty-two-year-old Matthew Boulton, is looking up at the skies ever so often. It doesn't look promising. It has been overcast since morning. The chief guest of the evening must return to London this very night no matter what. Matthew Boulton owns two large boats himself. In the early days, paddles were used to row these. But now, under advice from a confidant, the boats have been reconditioned. At first Boulton was extremely anxious with the changes made. Such big boats, how would they function without their trusty paddles? What if they broke down midway? But his friend had been adamant, 'If a man such as you is afraid of these engines, how would our countrymen believe in me, pray?' Indeed, true. The same friend had taken a patent some years ago for this very thing. Boulton has also jointly started a factory with this friend. Now the boats have pilots, not paddles. And last night they left for London to bring his guest over. Perhaps, they will manage to arrive before the club begins its proceedings tonight.

A big banquet has been planned for this evening, which accounts for the busyness around the estate. The staff has been particularly trained for such occasions where each month they must receive such honoured personages for a night's repast. The members are almost always the same and so the attendants are aware of their taste in food, where they wish to be seated, or what pleases their ears. It is the rare guest that they need to be ready for, so their master sends out his trusted men to gather whatever tidings they can as soon as their request has been accepted.

A carriage has halted outside their gates. Indeed, it is the first one they expect to arrive. The guard will now open the gates. Matthew Boulton will already be at the living room by the time his guest takes his first steps in, although his visitor dislikes being in the living room for long. He will head for the Soho House library almost at once.

A cool breeze floats in. It'll start raining any minute now. They must immediately make arrangements to get the fire going in the large dining hall. Matthew Boulton is walking down the stairs now. His friend has already walked into the living room. Servants stand alert just a distance away to take his coat, but they first await orders. However, their guest is staring grimly at the fire. Has something untoward taken place? Or has a lackey made a mistake?

Boulton coughs, gently. 'What is it, James? Why don't you hand over your coat?'

'Hmmm!' James Watt mumbles absentmindedly.

'You've come a long way in this chilly weather. Rest awhile. Look at the condition of your fingers.'

Finally, James Watt springs back to his senses. 'No, my friend, it is about a serious error on your part. How long do you expect to pull on like this?'

'What mistake? What did I do now?'

'Look how you are still using wood for fire. Can you not use coal? It's so easily available.'

'Ah! Now I understand. I'd completely forgotten about your profound coal-love.'

'You may well ignore it now, but there'll be a day when the whole of London will be under the spell of coal, at least coal engines, and that day, my friend, you'll think of this unfortunate soul who may not be alive to witness that sight.'

'Come now, utter balderdash! I've in fact just heard of a new wood type, mahogany. A brownish-red wood with no visible pores. It gleams upon being polished. Won't it be a fine idea to use it for our church? It's highly priced and can't be used in the fireplace though.'

'Don't mislead our conversation, Matthew. I wish to know why no coal is being used here for this fireplace?'

'My home is a faultless white. I do not wish it any dark spots.'

'Ah, I see.'

'Come now, take off your overcoat. Let's sit by the fire.'

James, his countenance serious, begins to unbutton his topcoat. The sound of horse hooves is heard.

'Someone's arrived. Must be Erasmus.' Matthew speaks up.

'It's not yet time for Erasmus. This must be Joseph.'

James is inordinately fond of Joseph, who has planned a surprise for everyone this evening. Erasmus Darwin is a good man, too. But then he is a doctor, and thus prone to being grave. James holds this large, heavy man in awe.

Matthew notices how James has fallen silent again. He seems frozen as he stares into the fire. The orderly is waiting by him, a glass of warm beverage in hand. He doesn't wish to break James's thoughts. The stillness will break soon enough with the arrival of the next guest. Matthew gestures that the attendant continue to wait.

The doors fly open and a booming heavy voice echoes all around, 'A very good afternoon, my dear gentlemen.'

Matthew's eyes shine joyfully. He was right. It's not Joseph, this is Erasmus Darwin.

Two servants appear instantly on both sides of Darwin. Erasmus is a large man, and his coats are styled differently. They will need a couple of hands to be taken off.

Erasmus stands parked like a monument. He has been working with trees all his life and now seems to be turning into an enormous one himself. Or so Matthew feels sometimes. His expressions are not visible in the dappled light of the fireplace. But his voice booms once again, 'What is the size of the earth, Matthew?'

'Why this sudden question?' Matthew counters.

'I've been thinking of this since morning. How large is our earth—in length, width, height. The world is not made up of England only, is it, that we can measure it quickly?'

'Why even speak of length and width? Haven't we understood it to be round?'

'We have, no doubt. But what if it is not? I say, can you take me outside of this world and show me how it appears from far? I cannot believe in anything till I have seen it with my own eyes. Let that be, I do not wish to know its shape, I want the measurements. You are digging away, cutting miles of canals from end to end. But ever thought of what remains hidden under the soil?'

Matthew steals a quick look at James Watt. Watt is still lost in thought. He is not a part of their conversation.

'One measures according to shape. One can't do it with strings, you know that well enough. It must be done mathematically. Different formulae for different shapes, circular or rectangular, it'll depend on that.'

'What say our respected pastors and reverends? Doesn't the Bible or Old Testament mention something?'

'Not much there. If God created the earth in six days…'

'Enough! I don't wish to hear those stories anymore. Tell me Matthew, with your hand on your heart, do you believe in this six-day theory? Robert Hooke, and even Nicolaus Steno, have shown us that many organisms inhabited the earth much before men did. These are now no doubt extinct. How are these related to the animals we see living now? In fact, who are we? Who was Adam, or Eve? Were they human beings? Where had they come from?'

'All our questions have answers in the Bible, Erasmus.'

'No, you will not find these answers there. If you consider the Bible infallible truth, then these creatures will have no identity at all, Matthew. They do not belong to the present. It was you who saw the evidence of their existence in the strange marks left on the surface of coal plates as the earth was being dug up. Imprints of leaves and flowers, too. How would they reach so deep into the earth otherwise?

Who could have sculpted those there? Hadn't you questioned, too?'

'True, I did say so. But it is best to set these questions aside now.'

'No. We will not sidestep them at all, sir. What kind of people would we be if we do not dig deep at truth?'

Josiah Wedgwood is standing at the door. Raindrops are dripping from his heavy long coat. He's of a tall frame, and the attendants seem undersized in comparison. They haven't quite managed to hold up the umbrella over his head. Next to him stands James's friend, Joseph Priestley.

James jumps up straight as soon as he spots Joseph. Josiah and Joseph greet him. James nods ever so slightly in acknowledgement and immediately clasps his hand upon Joseph's and pulls him inside.

'Listen to what Matthew is saying. Apparently, his rooms will darken with coal smoke. Look, Joseph, you were the one to separate oxygen from the gases, and you discovered carbon dioxide, too. Can you not separate the carbon from coal smoke?'

Matthew interjects. 'Why the rush? I promise to leave the destiny of coal to Joseph. Let him stop for a breath at least. Do not worry, we will all live a while longer on this earth.'

James feels suddenly embarrassed. He's surrounded by very respected citizens. He sits down again, suitably chastised like an errant schoolboy. A servant has brought warm water and warm drinks. In charming stoneware, all gifted by Wedgwood, from his factory. Pretty cups and bowls with tea twinkling in them. The first round is only tea. Boulton's guests like it that way. In another half hour the staff will begin to set down the dinner table. By then all the other respected visitors would have arrived even though the weather may not hold up. Each one of them is extremely particular about time.

Around two hours later, everyone sits down to dinner. It is exactly six o'clock. Their special guest has not yet arrived. It is around this massive unusually fashioned wooden dinner table that the gathering truly comes alive. A rectangular table, its corners rounded off. The master's chair at the head of the table, on which Boulton is sitting, is gilt topped. To his right is Erasmus and Wedgwood next. On the left is James with Joseph next to him. There's never any change in these five seats. The rest seat themselves as convenient. Joseph has excused himself now with Boulton's permission. The butler informs them that he's brought along a large box. He's to show them a trick even as dinner continues. Matthew Boulton does not know anything about this. Even James's question about what is to take place elicits only a smile.

Erasmus interjects, 'The Parliament has passed an Enclosure Act. Will the farmers benefit in any way?'

William Murdoch, better known as Gas-Light Murdoch, raises his hand. He has invented gaslights and set all of London alight. Matthew acknowledges him. Murdoch speaks up, 'Farmers will now be able to till the land as owners. Their land has been divided and marked carefully. Should they wish to, they can procure new machinery and grow more.'

'But the villages have not yet seen these changes, have they?'

'They will. Soon, and more. Wilkinson has thrown open the iron doors and made them privy to happy dreams. Everyone wants a piece of that heaven.'

'One change has taken place for sure,' intervenes William Small, 'the villages now have pedigreed goats and cows. No one eats salted meat in the winters anymore. They now eat fresh meat with bread.'

'But what of this canal business? The Duke of England wishes to cut canals across the length and breadth of the land,' questions Erasmus. He seeks the answer from a certain person.

Matthew sucks in his breath and turns towards James. But James is oblivious, busy lathering butter on his warm muffin. Tonight's special is roast lamb in orange and honey glaze. There is also a giant salmon baked with ginger and garlic. But no one has yet taken a piece. Dinner tonight features three wines—claret, sherry and madeira. The diners are silently savouring the deliciously spicy aroma of the lamb roast. There's a tuna-cherrycyder salad too, which a couple of the guests have relished. Only Erasmus has helped himself to the pottage of peas, beans, and onions. He's not too fond of meat, although he has served himself a couple of small pieces of rabbit. Tonight's menu also has a French dish on it—a pie, stuffed with pigeon meat.

Matthew notices Small shyly place two pie slices on his plate. The beef on the table is special, only the portion near the ribs has been cooked. The smallest ribs are the softest. These pieces have been soaked in wine for half hour or so, then slow roasted on fire, one side at a time, allowing the fat to meld with the spices well. Priestley's eyes are shut tight, as he chews just such a piece, winging his way through the flavours of a different universe.

Matthew answers now. 'Where's the harm? It would indeed be great if the waterways could be made worthy of use, right from Carlisle to Cornwall, from Dover to Dumfries, and across. What would be better than joining all the waters possible, Thames, Severn, Mersey, Trent?'

'Better for who?' queries Erasmus again. 'Land is being taken away from the farmers by force, people are being rooted out of their homes. Are you aware of what is happening in Somerset? A beautiful green village has been turned into a ditch. The whole town looks like a dug-up trench.'

Small adds, 'Have you read Evangelist Wesley's statement? Every place on this earth can be inhabited and lived in. The only one that is out of bounds is the town of Midsomer Norton in Somerset. It resembles a dug-up hell now.'

'Indeed, the Parliament is clearing one Canal Act after the other. And Navigation Acts to match pace, too. They will destroy this country for sure. People are digging into the grounds below their homes. I suppose everyone thinks they'll become a coal mine owner,' Erasmus adds again.

'If only it was as easy to become a mine owner,' counters Priestley. 'Surely, we've all heard of the two farmers who took their own lives in Camerton recently. They'd sold everything they owned—land, cows—just to dig under. And what did they come up with? Not coal, just black stones. Completely non-flammable like granite.'

'One cannot shut down all enterprise on the basis of a stray case or two,' refutes Matthew.

All at once everyone falls silent.

Finally, Josiah Wedgwood breaks the stillness. 'We cannot depend solely on Somerset coal. It has served us for long but is now no longer able to. Coal cannot be found horizontally across the ground, it's more like a folded carpet. A large portion of it is deep inside the soil.'

'What a blasted problem,' says Murdoch Westphalia, 'in the west of England the layers of coal are thick, almost like a sandwich between layers of sandstones. And yet the situation in Wales, Bath, even Somerset is so different, almost as if someone has pressed these into folds. And large cracks can be noticed in layer after layer. How has all this happened?'

'There is much we cannot understand,' says Erasmus, his face serious, 'the Bible does not hold answers to these doubts.'

'Fie, let all that be,' Matthew says immediately, 'why must we drag in religion everywhere.'

'There is no tug of war here with religion, Matthew, it is where it should be. Regardless, the need of the hour is to have a detailed map. There is no doubt we know our geography. We have drawn maps of our roads and waterways. So, it is not geographical maps that we need, but maps of the underground rock strata to know what lies beneath the ground.'

James suddenly looks up, taken aback, 'A map for layers below the earth? Can that be done?'

'Mr Watt, who knows better than you that there is no word such as impossible. A map of the earth's crust. How are all the layers arranged? Where are the cracks, where does a layer end and how? Do you not think that it will be easier to extract coal if we know these facts better?'

By now Priestley has once again sought his host's permission and stepped inside.

'Indeed, I do. I think of it often. But how will we even find out what lies below?'

'Why can't we? After all, aren't we digging up the entire countryside for our canals, and exposing all the rock layers below?' This was James, in sudden support of Erasmus, a man he often argued with.

'That may be, but how does one know what lies six miles deep within, maybe seven miles even? White, black, maybe yellow, and red too, so many shades and types! If the world was created in six days, do we then take it that all the layers under the surface across the world would also be the same?'

Matthew interjects, 'Before we are deflected from our conversation, I'd like to inform everyone that the respected Charles Clomore has donated a substantial portion of his Birmingham and Newhall estates to build a new church. Work will commence early next year. James and I will also donate two pews engraved with our names. I heartily invite our Lunaticks congregation to join us there. You are most welcome to use any one of the two rather than join the crowd.'

William Withering is delighted, 'It'll be a very big church, I hear. I believe its spire can be seen from London too.'

'Have you decided on its name?' queries Jonathan Stokes. Stokes has just completed his dinner. He's particular about not speaking while supping.

'Hmmm, yes, I have, but am sharing this in confidence right now. I do hope that it'll not be shared outside the fourteen of us, at least not till a month has gone by after the foundation is laid. What is it, James, do you recall?'

'Why are you being so secretive? Is it a combination of both your names, the Boult-Watt Church,' Erasmus asks, his eyes twinkling naughtily.

'How could you even say this? We are inconsequential to this. The church is named to honour Saint Paul—Saint Paul's Cathedral.'

The butler is now serving an array of desserts in tiny silver bowls—delicious strawberry tart decorated with bits of ginger soaked in honey and wine, tarte aux fraises with chopped ginger, ravioli bursting with fruits, strawberry pudding, almond pudding, squab pudding with berry sauce—an astounding range in keeping with the host's immense love for varied desserts.

For the lovers of custard, there's sweetened cream almayne dunked with chopped apples as well as a rainbow dariole bursting with colours.

Each member of the group is picking up their favoured dessert bowl. Just then Priestley comes bursting in, 'I beg your pardon, my respected friends. I'd stepped out with permission from our host. In truth I wish to astound the members of our group today with an amazing creation. It's a new drink that I've made. Please do try it before you taste your desserts. Once you've downed this, you'll feel lighter, hungry even. You could easily have your puddings a couple of hours later, please.'

'I thought you wished to surprise us, Joseph,' Watts counters, a grimace plastered on his face.

'This is the surprise.'

'This, a surprise? You've learnt a new dish, and we need to swallow it for your sake! Oh, I see now, dearest Joseph Priestley, Fellow of the Royal Society, has dumped his scientific temper and taken to a passion for culinary delights. A surprise, indeed.'

'This is not mere cooking, my dear fellow. I beseech you, my dear host, Sir Matthew, please do partake of this.'

It's a slim glass tube, with a deep wine-coloured liquid inside. Matthew hesitates. 'Do I need to drink this? Why not let James try it instead?'

James Watt immediately interjects. 'No, not at all. How can that be? Indeed, the host in whose home this has been made must take of it.'

The members are flabbergasted. Are they now supposed to swallow some outlandish liquid? And none knows what its outcome will be. Damn Priestley, could he have found no other place to test this? It's pouring outside. No one can even escape, no excuse will work in this weather.

Priestley's face is now ashen. He's standing frozen, the tube in his hand. He could himself drink it up, but then no one will know of its worth. Do these civilized folk not trust him at all? Do they think he's feeding them poison? Well, so be it. He'll drink it up himself.

'If our host permits, I could take a swig from this.' Each pair of eyes turns in swift surprise towards this heavy voice. Erasmus Darwin! His face is serious, but his eyes glitter as he stares at Matthew.

Boulton breathes easy suddenly. Garrulous with relief, he says, 'Of course, indeed, go ahead. It takes a scientist to establish the credentials of another scientist, of course. You have my wholehearted support, Mr Darwin.'

Priestley cuts in again, 'This is not to be tasted, sir, it must be had entirely.'

'Fine, let me have it then.'

The tube is passed on. Erasmus steals a quick look at the cross nailed on the wall in front. Then he empties the tube in one gulp. Priestley says again, 'Not so fast, sir, take it slowly.'

Erasmus hears nothing. But a large burp escapes him once he is done. His eyes tear up. Burping in public is such a shameful act. But nobody is paying any attention to that right now. Matthew Boulton has stood up, 'Erasmus,' he asks urgently, 'are you fine? Just indicate if you are feeling unwell. You are a doctor yourself, so tell. Dear God, help us.'

Erasmus is sitting still, his head hanging down on his chest. Servants have come running at Matthew Boulton's frightened cries. They are awaiting orders to fetch a doctor.

Finally, Erasmus lifts his head. His eyes are still smarting, but is that a hint of a smile across his lips? Matthew Boulton is truly alarmed. He questions again, 'What is it, my friend, what's troubling you?'

'What a brilliant thing this was, truly it was. I feel so light, as if I'd barely eaten a thing. All the heaviness has escaped from my mouth. How did you make this digestive, my good man?'

'With carbon dioxide.'

'What?' Erasmus still looks stunned.

'Nothing that should worry you, it is harmless. This was a carbonated drink.'

'I'm delighted, Priestley. We needed just this thing.'

'This surely means another patent, what say?' adds James, his face suffused with unfeigned thrill.

'Let me add a note of joy to this joyous moment,' Erasmus smiles. 'James Watt had been trying to discover coal that wouldn't blacken anything around it, and we have discovered it at last. Let Watt now take up the challenge to improve on this further. This will be discussed in our next meeting.'

A servant whispers something into Boulton's ears. And immediately comes Matthew's pronouncement, 'With great pleasure I announce the arrival of our special guest tonight, an engineer who has designed several bridges, canals, harbours and is currently working on a lighthouse. He has pioneered the use of hydraulic lime in concrete, using pebbles and powdered brick as aggregate. Would our members be able to recognise this gentleman?'

Josiah Wedgwood raises his hand, 'FRS John Smeaton.'

'Absolutely,' Matthew beams, 'Let's all head to the living room now, and relish his amazing stories of discovery as we enjoy our tea.'

YEAR 1787, OXFORDSHIRE

The boy is standing in front of his uncle. His head is weighed down to his knees in embarrassment. He's just turned eighteen. He had lost his father as a child and was forced to take shelter at his uncle's along with his three brothers and a sister. They were all brought up here. Within six months of his father's passing, his mother had married again. At first, she would take the children with her on occasion. But her circumstances had changed with time. She couldn't manage beyond her newly acquired husband and the newborn she had with him. Their new stepfather had become crabbier with each passing day. He would beat his wife up at the slightest mistake. And she, who couldn't protest, instead became angrier with them. This, of course, is the story in every village, a situation that unfolds in every home. At first, no one was bothered. But soon, things got out of hand. Maddened by hunger, one day the little sister began suckling at the cow's udders. She was beaten up with a metal baton and left for dead. It was the villagers who saved her by bringing in a doctor from the neighbouring village. But it had still not made a jot of difference to their mother. All she said was, 'When she has done wrong, she must be punished.'

That day the boy had fallen at the feet of the local priest and wept bitterly. He was the oldest of his siblings. Every time they sobbed, it was he they clung to. But he himself was only ten. It was not yet his time to earn and feed them.

He confided, 'My uncle lives in the next village. He too is called William. He is William Sr., and I, William Jr., I beg of you to send us to our uncle.'

The priest knows William Sr. He works his own land unlike his brother John Blacksmith, who worked with metal. William Sr. has come into several acres of land after the Enclosure Act was passed by the Parliament.

The priest had called together the village elders. 'What do you know of William Sr.'s condition?'

The unanimous reply was, 'He's doing well for himself. He has the means to get modern equipment from town for his fields. His fields yield crops at least three times every year. He makes money on the side too, by loaning his machines to other farmers.'

'It seems then that he could take up the upkeep of his nephews and niece.'

'That he can, but the man's a miser, and a moneygrubber at that! He's not going to give in that easily.'

The priest went about collecting more information. William Sr. owned a dairy farm in addition to land where he kept prized Hereford cattle especially bought from Scotland. His farm boasted at least thirteen excellent Southdown sheep.

A meeting was called. William Sr. was summoned. Anne, the wife of the deceased John, was seen too, a tiny one-year old tucked in her arms. As were present John Blacksmith's four children—William, John, Daniel and Elizabeth. The village venerables had all been requested to join as well.

The village headman wanted to know if William Sr. was aware of the plight of the Blacksmith brood. William Sr.'s countenance immediately reflected the right amount of sorrow. Yes, he knew all about them, but was right honest confused about the correct way forward.

The priest immediately addressed the situation without preamble. 'Look here, William, we are aware that you are a respected farmer in the neighbourhood. It is not at all fair that your nephews and niece should die of starvation while you live as well as you do. Whatever the circumstances, they are your blood.'

The mention of these heirs melted William a little. He suddenly saw the three boys, like cherubs they looked, and one tiny girl. All four pairs of eyes were looking upon him beseechingly, teary-eyed. These were his brother John's offspring. The news of Anne's marriage after his brother's death had reached him indeed. But he had never checked upon these children, he had not wished to. He didn't care a jot at being called 'Stingy Willie' by the villagers. Had he not known what money truly meant, how would he, a man with a sliver of land to till, have become almost a landowner? Now, he is well-known in his church. He's purchased a congregation bench for himself. The villagers pay to sit on it!

William Sr. declared, 'Yes, indeed, but to take on the upkeep of four of them—just pause a minute, Father, I cannot keep them like cattle, can I? I must look after them as well as I do my own.'

The priest sighed. He'd known how this man would gripe. A miser from head to toe, a man who doesn't even keep his own brood with clean clothes. The priest had already discussed the entire matter with the archbishop in advance.

He offered, 'William, from next month on, you need give only half the money you do every month for the upkeep of the church. With the other half, look after these unfortunates. Our benevolent Archbishop desires this. If you refuse, we will hand over their responsibility to someone else. But it is they who will then benefit from the arrangement.'

'No, not at all, how can you even suggest this? I cannot think of sending my brother's children to another's care. I wish to take them with me right away, Father.'

The priest quickly hid a smile. Village folks hand over nine per cent of their earnings to the church. Those like William Sr., prosperous and comfortable, would naturally be giving away a substantial sum. He'd known that the parsimonious man would take the bait.

William Sr. stared fixedly at William Jr., the oldest of the lot. John's other kids still seemed very young. Only this boy is on the threshold of youth. Even if not in the fields, for sure the lad could be put to work at the dairy farm.

William Jr. clutched his siblings and came to stand by his uncle. The older man asked, 'What of your belongings? No clothes or stuff?'

The boy replied stiffly, 'We have nothing more.'

Anne turned her head away from them all. The older man wasn't too pleased with the young one's manner. Cheeky imp. But he held his tongue, lest the priest reprove him again. He said quickly, 'Never mind, let's get going. We'll figure out the rest soon.'

William Jr. walked up to the priest. The pastor drew him close and gently kissed his forehead. He whispered, 'May the Lord God look after you. Take good care of your brothers and sister.'

A boy, not yet ten, and a face as hard as stone. Blue veins stood out just below his cheekbones. The priest couldn't bring himself to say anymore, not even to prod him to bid his mother goodbye. The boy looked at no one. Anne gazed at him powerlessly. And he gave her no chance to indulge in any emotional spectacle. He simply took his siblings by hand and began to walk away.

It's now eight years since that eventful day, the very day that William Sr. said to his nephew as soon as they stepped into his home, 'I cannot feed idle mouths. You work, you eat. Your uncle has to work hard too to feed himself.'

William Jr. was calm, 'I'll do as you instruct. I can start now. But I have two conditions.'

A fatherless boy, cast off by his mother, with no clue about his next meal or bed, and here's he, laying conditions!

The senior man held himself back, or else he would have yelled his lungs out. This boy was nothing like any other child that age, not simple, certainly not soft. Best to avoid trouble, who knows he might just complain directly to the archbishop. All said and done, William Sr. had a social reputation to maintain. If he wished to continue in the village, he must follow the dictates of the church.

William Sr. waited, goggle-eyed, just as he did every time he was on the verge of picking up a fight with a customer over milk. He asked gruffly, 'What conditions?'

'My siblings are very young, you cannot put them to labour, not in the field, not at home.'

'Hmmm, fine. They are young, no doubt. And, the second?'

'You have to let me study.'

'And how will you do that? No one in this village does it. Those who do, have left for the city. There's only one school here, and one master. I've heard he sits all day, a cat on his lap. Billie, that's his name.'

The boy stopped him right there, 'I'll study with Billie then.'

'But I can't pay for it.'

'You don't have to pay the master, and I'll manage my books. Only pay if I need a few special ones.'

William Sr. didn't have a clue about the cost of books. And he wouldn't pay for it all. But he did need someone he could trust with his dairy farm. This boy didn't strike him as a thief. 'Fine,' he had conceded, 'start work tomorrow. At dawn, at four, you need to milk the cows. As for the rest, I'll tell you by and by.'

'You'll have to let me free while school's on.'

'Fine, fine, we'll see. The blacksmith's boy wants to be a master. Now get going, your aunt will show you your room. It's next to the granary. Take your siblings and settle down.'

***

That bit of a boy has now turned eighteen. He studies even today. There are no classrooms for Billie's students; he teaches when he can. There are no promotions, no degrees, no nothing. No exams either. And yet people say that William Jr. has learnt all of Billie's books by heart. He will make the village proud. He has no mind for work, wanders around all day, knows nothing about farming, and God knows what people see in him that they praise him to the skies. He's always orbiting around some weird men. Just last year, a mad-looking man, supposedly the repository of knowledge and wisdom, had stopped by. He had polished off a full mug of coffee and sung high praise for the boy, 'Your nephew is an extraordinary one. Watch my words, he'll be someone to reckon with someday. He just needs a chance.'

William Sr. is exasperated. 'How many more chances should I offer, tell me that? He lives free, eats free. He's now a youth and yet no good at the farm. I put him in the dairy and there too he created trouble.'

'Created trouble. Oh dear, no! The weighing stone in your dairy is not a mere stone but a fossil, and it is this boy who told me of it.'

'A fossil, what's that?'

The gentleman said with great interest. 'It is the remains of a living thing. The poundstones all look the same, round and flat with markings on them. Almost as if you have flattened an orange.'

'Enough, I don't wish to hear more. I've seen these since birth. How does it matter?'

'Matter? It means the world, William, my man. I've seen living sea urchins at the shores, much like these. The poundstones are predecessors to these creatures, they were birthed long before. The point to note is that these stones are now being found in Churchill and its neighbourhood, places so far from the sea and its shores. Dead sea creatures cannot move so far on their own, so did someone carry them here?'

'Who would have carted them here? Haven't you heard of the great flood, and of Noah? The flood must have washed these up here.'

'Quite right, my man, you are on the right track. But not about Noah's flood. Had it been such a flood, these would be spread across the land. But not as they are now, found within a short area, layers packed between sandstone. As your nephew says, this place was sea and is not sea anymore. His argument has weight. What do you think?'

William Sr. suddenly began yelling, 'There's a limit to tolerating nonsense, sir. You've been spewing gibberish since you have arrived—poundstone, fossil, sea, god knows what! What use would all this be, whatever it is you are trying to unearth by shaking up everything possible. What would a farmer like me gain by learning of the poundstone's history? Will the cows start producing more milk if I feed them your thesis? It's company like yours that has turned my nephew into a crazed incompetent lad, rambling around all day, scribbling away.'

'It's not scribbling, it's called observation. The second rung in scientific work. Occurrence, observation, testing and then collecting facts. Your nephew is keeping records of it all. The modern equipment that you use today, those too are the gifts of science.'

'I do not understand your science. He's a blacksmith's son. If he wishes to forge useful farming tools, let him do so. I'll get him what he needs. All this is pretend-work to me, this writing-scribbling but no shaping things with a hammer.'

'It's not my task to make you understand, William. All I say is let the boy do as he finds fit. Stand by him. After all, you are his kin. One day, this boy will stand with you, and the world will shower praise on you too on his account.'

William Sr.'s behaviour did not show any change even after this conversation. Of course, he's not one to use violence. And in any case, he cannot beat up such a grown boy. But he is now inspecting the accounts that are totting up against the boy's name. He knows he cannot put the boy to work in the farm.

***

The boy is still waiting. What a whine he's turned out to be, always asking for more.

'What do you want?' asks thc oldcr William, his voicc almost a growl.

'I want to buy a book.'

'What sort of a book?'

'One that teaches calculations.'

'Calculations?' A hint of urgency has crept into the senior man's voice. 'Can you learn to make calculations of our land?'

'Yes, indeed. And soon, the layers below the earth…'

'Why in the name of heaven are you so keen on what lies below? Why can't your work be above the earth? Why not focus instead on wheat farming, or corn, or apple orchards, even grape yards?'

'Yes, that can be done.'

'Does your book have all this?'

'Perhaps.'

'That means you don't know what it says.'

'How would I? Have I seen or read the book? But I've heard.'

'So far whenever I've given you money for books…'

'Thrice so far.'

'What did you say?'

'You've given me money thrice. Three shillings at first, then seven, and four shillings the last time.'

'You're a fine one for keeping tally, I see. So, what did you buy with so much money?'

'That's not enough to buy, uncle. But I used it to rent and read books at the bookshop.'

'That's not so bad. The shop's right in the next village.'

'The shop in the next village sells crockery and only books to learn the English alphabet.'

'So that's not good enough.'

William Jr. gave his uncle a piercing look, 'I'm talking of books on science and engineering. As for the alphabets, I've already studied all the books.'

William Sr. does not know what to say anymore. He has no clue if there are books other than those with alphabets. Finally, he says, 'Let that be, how much will this book cost you?'

'A half pound.'

'Dear Lord! That's a big lot of alphabets. What will you do after reading so many?'

'It's not spellings, uncle, it's measurements.'

'It's all the same. Who needs a book for measurements! Old George of our village, the one who's put four wives to rest and is not called Dead-wife George for nothing, does all his measurements, all in the head, without having ever read a word. But I'll give you the money since you've asked for it, after all you are my scholar nephew. Get your aunt. My dear, give the boy five shillings.'

'The book's worth a half pound, uncle.'

'Never mind, just haggle a little. If needed, just labour a bit for wages at the shopkeeper's. Where is the shop by the way?'

'Oxford.'

'Don't say! You have to travel twenty or more miles to get to Oxford, and all for a book? I can't pay your fare, I hope you know that.'

'I'll walk. I'll be back by tomorrow morning.'

A young boy, walking across a dense forest in the middle of the night was an alarming thought. 'Go then if you wish to, so stubborn, you won't listen to reason at all.'

'Please give me the entire amount, uncle. You can take it from my wages.'

'What? Are you labour at my farm that I pay you wages? What'll folks say if they heard you? Okay, let that be. Here's three shillings more. Now don't you whine any more. Eight shillings should be enough to fetch you the book.'

William is sitting in his room next to the barn, a tiny space, overpoweringly musty. There's dung wherever the eye goes and flies buzz around. A lone window. William Sr.'s house has imprisoned the sun. The boy's counting his money. He has four shillings saved up. If he borrows from Billie, how much would that be. At most two shillings, if at all. He wants to buy the book, not rent it. In any case it'll take him more than an evening to read it. Does he have to beg for this book?

He comes to sit by his little sister Elizabeth. She's barely recovered from a fever, her face is ashen still.

The boy begins, 'The priest had come to see you, do you remember, Liz?'

The girl nods.

'He gave you two shillings then, right? Have you spent it all?'

Elizabeth is silent.

'My darling sister. Will you give it to me? I promise I'll return it to you.'

Elizabeth shakes her head vigourously.

'Please, sister, you know that I teach Billie's students. Billie pays me a shilling or two for this sometimes. I'll give it to you next time. I really need the money.'

Elizabeth tugs hard at his hand, and points under the collapsing wooden bed-plank. William spots a pile of tattered cloth.

'What's there?'

'Look under.'

Two small clay penny pots hidden under the rags!

'Whose are these?'

'Dan and John. They put in whatever they get during Christmas.'

William picks up the pots, 'I'm not taking yours. But you've kept it safely, right? Show me where you've hidden it, just so we make sure they don't pinch yours.'

Elizabeth silently points to a gap in the wall. Her little pot is behind a broken comb.

'Oh no, Liz, that's an unsafe spot. The boys will take it away if they see it. Don't you worry, I'll keep it hidden in the barn house, behind Rosie's trough where no one will notice. Your favourite cow, the red Rosie. You love her, don't you?'

Elizabeth nods again.

'Isn't this a great plan?'

Liz claps happily.

'I'll see you soon. Just go off to sleep and don't wander off alone to check on Rosie. Call me when you want, I'll take you to her.'

Elizabeth pulls up her cover and dozes off. William breaks the two penny pots. From Daniel's he retrieves three shillings, and five from John's. Elizabeth's is the best, seven shillings in total. She is the youngest of them all, so she gets the most too. William has twenty-one shillings in all now. How will he manage nine more? Should he offer his labour for wages in the houses on his way to Oxford? He must leave home before Dan and John return from school. He'll see what follows.

The next afternoon the villagers notice John Blacksmith's eldest son William walking back from somewhere, a book tucked under his arms. William heads straight to Billie who had lent two shillings to him. The Master is educated but curiously uncurious about everything.

Billie yawns as he enquires, 'Got your book?'

William's eyes gleam with pleasure through his back-breaking tired expression. Billie perks up a little, 'What book is this?'

'*The Art of Measuring*.'

'Pfftt! Couldn't you get some juicy romances? Will you check if there are books on cats, eh?'

'Why not? We could both go back together someday.'

'No way, it's too far.'

Billie quickly shut his eyes. He often falls asleep mid-sentence. William picks up his book and begins walking towards the door.

'Hey,' Billie wakes up suddenly, 'had to give you some news.'

William looks back.

'Have you heard of Edward Webb?'

'The great surveyor! He who is surveying most of the roads, canals and rivers of England?'

'Yes, the same. He is coming to our village soon. Some farmers from Churchill have hired him to survey their land.'

'Don't say! When?'

'Hold your horses, there's more. That well-known gentleman from Churchill, the one who travelled to India. What's his name, I'm forgetting?'

'Hastings. Sir Warren Hastings.'

'I hear he's coming back home. He's made himself a lot of money, but picked up a terrible reputation too. D'you know that? He's accused of corruption, murder, killings, brutality. He'll be tried once he gets back. Anyhow, he's contracted Mr Webb as a surveyor. What happened? Why did you jump up like that? You're unbelievable. Where do you find this energy? After a long night of no sleep, no food and walking all that way and back! Be careful. I can only hope that there's food for you at home.'

YEAR 1787, NORFOLK

It's Lord Denver's cake cutting ceremony for his twenty-fifth wedding anniversary. Amongst the invitees are notable personalities such as Lord Vernon and Lord Digby who have arrived accompanied by their families. Since the Lord Lieutenant Carmarthenshire is busy at work, his wife, the Lady Carmarthenshire is attending the party by herself. Parties hosted by barons are usually attended by barons. But, in an exception, the special invitee to this party is the Duke of Manchester. The cake cutting will commence only after his arrival. A pastelero has been brought in from Paris for the cake. A four-tiered, almost five feet tall cake, with a bunch of wine-coloured roses carved in cream lining one of its sides. The top tier boasts a gold crown with the royal Denver emblem.

The Baron and Baroness Denver are both over six feet tall. They could well stand on their own feet to cut the cake. But there is still a hitch. They can't place the cake on grass. So, a rosewood tray table has been specially ordered for it. It's a delight, the table, legs gleaming silver, the top worked with copper and gold. When the cake is placed on it, it'll be over seven feet in height. This will mean that the Lord and Lady Denver must stand atop a tall stool to cut it.

This is precisely the problem. It was Lord Denver's desire to invite only his compeers and those junior to him in rank. Only the barons, baronet, knights and esquires.[1] But Lady Denver has kicked up a fuss. On her father's side she belongs to the Earl of Talbot's family. And the Duke of Manchester is a dear friend of the Earl of Talbot. As a result, every duke and earl in the county has now been sent an invite. Indeed, she herself has had to carry the invites for them. In her mind she was sure that none of these invitees would deign to arrive, since it is socially unusual for high-ranking families to show up to these events, unless they are closely related or friends. Who would have imagined that at least two such families would indeed participate!

Now, if the Denvers stand on a high stool in the presence of these high-standing guests, it would go against all social norms. The banquet hall is not a theatre gallery where this would go unnoticed. The time decided for the cake cutting

1 According to rank, the King is at the top, and the duke below him. Then, in order, come the marquis, earl, viscount, baron, baronet, knight, esquire and finally, the commoners, the gentlemen.

is thirty minutes past noon. Right now, it's showing just past twelve in the pendulum clock. The baroness has not yet dressed up for the event.

The Chief Assistant will ring the warning bell. At the third call the trumpets will sound. The Denvers will sashay down the stairs, hand in hand. As for the rest, Lord Denver is drawing a complete blank. If he stands on the ground, it'll be impossible to cut the cake. It'll become a matter of ridicule. He cannot even begin to imagine the situation. And if he stands on the podium, it'll be a matter of utter shame.

Lord Denver is pacing agitatedly in his dressing room. The baroness is in her fitting room, right next to his. His door opens a little, and then she peeps in, a big smile on her face. Their attendants exit immediately, as per protocol.

'How do I look?' she preens.

Lord Denver answers absent-mindedly, his voice troubled. 'Quite nice!'

'Just nice? You haven't even noticed my diamond necklace.'

'Oh yes. It's stunning.' But his voice has no shine whatsoever.

'I know, but Dorothy said these diamonds are not from India, they're from South Africa. The cuts are not too well-defined. And these are rose pink, not clear.'

'Really?'

'What are you thinking? Even on this special day I'm not able to draw your attention.'

'Not at all, I'm all ears. Who's Dorothy?'

'Why are we talking of Dorothy?'

'Because you mentioned how Dorothy got you the diamonds.'

'Dorothy helps me with my jewels. Why would she get me diamonds? You got these for me. Our friend in India, Richardson, sent these. You've spent seventeen thousand pounds for them. At least that's what you said. I hope it is the truth, or is it not?'

'Yes, of course, that's correct. Lovely diamonds, truly amazing.'

'You're not listening to a word I'm saying. You haven't even looked at me properly.' Lady Denver pouts, her voice all weepy.

'What gave you the impression? I'm in a spot of trouble, Liza. You made a huge mistake without meaning to.'

'Tell me now! How?'

'Where was the need to invite earls and dukes? It's created untold troubles.'

'Oh dear, really? I've never heard that the presence of earls and dukes could ever create trouble! Did you see an earl or a duke at Lord Digby's twenty-fourth marriage anniversary last year? Or even an earl perhaps at Lord Yorkshire's dog Lucia's birthday? Forget the dukes, these people have accepted our invite simply because we are related. Thanks really to my father's renown.'

And thus begins her litany to her paternal lineage. If only the woman had an iota of intelligence to grasp the gravity of the situation. And now I must go through the torture of this blather for a bit, thinks Lord Denver.

He intervenes, 'Darling, how right you are, you the daughter of such an incredible family. Go now, finish your toilette. Your left eye needs a bit of colour, just a touch. And a dash of rouge too, please.'

'Oh dear, silly me! I forgot to pay attention to myself in the rush to show off my diamonds. I'll take your leave now. But do please not disturb me again, you will remember that my heart, won't you?'

Lord Denver forces a laugh out of himself. There is still no news of Knight Henry. He trusts the young man immensely. He's shared his nervousness only with this lad. Perhaps, he will find a way out. Lord Denver looks at his watch.

Guests have begun to gather in the hall downstairs. But there is no duke or earl among them yet. The royal drawing room will seat the special guests when they arrive. Downstairs though, a few women are grouped together, chattering away, the wives of barons and baronets. The wives of the knights are huddled elsewhere. They are all discussing the next play at the London Royal. These women are related to each other, cousins from the same family. They call each other by first names when alone.

'They are showing *The Rivals* at the Royal next Wednesday. Will you be attending, Lydia?' asks Isabell.

'*The Rivals*, a new comedy, is it?' Lydia queries back.

'Well, that's what I heard. I want to see Mr Woodward acting. It must be a fantastic play, particularly with him in it.'

Lydia answers, 'I fancy historical plays though. And it'll be *Prometheus* staged the Friday after. The male lead will be Rainer. Truly, how I adore that man. When he stands up, bare chested, a sword held across him, he looks a veritable Greek god, like Hercules.'

'Oh, have you seen Hercules yourself, Lydia?' Matilda butts in. Isabell steals a glance at her, lips curled in a smile.

'No, I haven't, but what is to stop me from imagining him? Fair, a chest full of dark curls, his arms rippling. I dream him up as I wish!'

'Speak no more, I understand it all. But do keep your wits about you, Lydia. You will be in a spot should you be screaming out Rainer's name while jumping on the baron in bed at night!'

'We are all one better than the other at being careful, dear Isabell,' Lydia's eyes glint naughtily, 'but this play is more on my list because of the songs by Renfold and Bellamy. The dancer is Signora Valdini. In fact, I completely forgot to tell you all, I'm taking dance lessons with her.'

Isabell stares at Lydia's fleshy frame and exclaims, 'Really, you've brought a stage artist straight into your home! Careful dear, do not let Valdini's waxlike smoothness melt others at home.'

'This is why I don't share any information with any of you. Always so negative.'

'Why are you annoyed? I was just teasing. Has the baron seen your dance?'

'Not yet. Let me learn properly first. I'll invite you all too one day.'

'Do that,' says Ruth, 'but don't bring your dance teacher that day. My lord goes to the plays only to savour these dancing women. Even if sitting far off, he looks ready to devour them at the slightest opportunity.'

'You are right, men are mad for Valdini.' This is Matilda.

'It's really his business what he does behind closed doors, but that he should salivate in public, I can't bear the thought.'

'You'll go for *The Rivals*, right, Matilda?' asks Isabell.

'Perhaps. But I too adore historical plays like Lydia. I've watched Sheridan's *Hamlet* thrice over. How I drool over him. That smooth candle-like neck, and that flaming glow of golden hair. I so do wish to meet him in person.'

'But you do see him, don't you, whenever you and the baron hand over a prize to him,' counters Lydia.

'That's not the meeting she hopes for. It must be in private, at home. What say, Lydia?' responds Ruth. 'And what of his attire? Would he even be in attire, or perhaps Sheridan will be without?'

'Shush, shameless hussy! You say whatever comes to your mind. We are burning with desire, and this girl goes about stoking the fire some more.'

'Those of you who'll miss the Wednesday play at the Royal, will lose a huge opportunity,' says Isabell.

'A huge opportunity? What are you saying? Are you expecting a large group of prospective sons-in-law there?'

'Well, one prospective son-in-law certainly, but depends on whose luck favours them. The Duke of Bridgewater, Francis Egerton, it is.'

'Pfft, what nonsense. I've seen the theatre card, and it doesn't mention a single duke.'

'It's all hush-hush. Perhaps he too doesn't wish the matter to be known. The duke's personal secretary is my husband's cousin from his mother's side. My lord mentioned it the other day at the dinner table. It is not to be talked about, my dears. Lydia particularly can never keep a secret.'

'As if you all can,' Lydia pouts. 'And what of all the scandals about the families of dukes and barons? Those come from you, Isabell.'

'Come now, stop bickering,' intervenes Ruth. 'Is the news true, Isabell? Is the Duke of Bridgewater truly keen on the play? Such a humourless man that!'

'Not simply humourless, quite an oaf. The man doesn't know how to respect women.'

'I've heard that too. I believe he considers coal more profitable, far more useful, than women. Can you believe it, comparing women to coal!'

'The man's gone bonkers about coal,' Lydia says, 'I've heard of his lack of civility. He blows smoke into faces while pulling on his pipe, speaks crudely without any concern for others. He can't stand women, too.'

'He's been a horror since childhood. It seems he was so foolish when young that his father was sure that the title of a duke should be denied to him. But there wasn't much he could do when his eldest son passed away untimely. The man is as unlettered as he is savage.'

'I've heard he's very passionate about art these days. He's picking up some very fine pieces.'

'Passion my foot! Wait and watch, there must be a reason behind this too. He has only two obsessions, other than coal. Races and canals.'

'Oh, that reminds me, my lord mentioned that the duke dreams of populating the entire country with canals. To transport coal. Dear mother of mine, a dark product no less, and so much song and dance about it. I'd see some sense if he was behaving like this about flowers.'

'Oh dear, seems you haven't heard! The baron has already warned the earls that they are not to take flowers to him ever. He's uprooted all the flowering plants in his garden.'

'So, are they to take coal as gifts?'

'How vulgar. I'd picked up a piece once, and my hand was stained black. Just like the faces of our serfs. This is all extremely plebeian.'

'Well, the duke is exactly that.'

'Now leave it be. No point talking about it. Who knows, somebody might overhear something. They'll just speak badly about us.'

'It's so like all of you to digress from the main topic at hand to irrelevant subjects. What if we were to present our daughters to the Duke of Bridgewater?' queries Isabell.

'What are you even suggesting! An irascible old duke, and we are to give our daughters to that barbarian?' Lydia rolls up her eyes.

'I agree completely. But he is single. The girl who marries him will become the Duchess of Bridgewater, assured of a lifetime of honour,' counters Isabell.

'Who will you take? After all, both your daughters are of age.'

'Not Clara. I've already lined up someone for her. I'll take Sophia.'

'Ah ha, Sophia! That tender, tulip-like girl. One can barely hear her when she opens her pretty mouth. And you wish to sacrifice her to that demon? Can you, as a mother?' questions Ruth.

'If our parents could, so can we,' counters Isabell.

Everyone is silent for a while. No one dares say anything after what Isabell has said.

Ruth speaks up again in a bit. 'Be that as it may, but a man about whom we all know so much?'

'Some take poison willfully, some in ignorance. We've heard stories about him, after all. What if he is different? And if there are tales, they are bound to be tall. We know how easily people can transform from human to demon within the walls of their homes.'

'Well, let me see if I can get Emma to agree to come too,' joins Lydia, 'I'll need to arrange for three guest cards now. It'll need to be done cleverly, so Emma and Sophia can be seated next to each other. Will you join us, Ruth, since Claudia is also of age now?'

'That girl's so bashful, she'll never join us if she's told the real reason. Let me try nevertheless.'

Isabell doesn't bother to check with Matilda at all. Her expression is deterrent enough.

The trumpets are playing now. Lord and Lady Denver are walking down the carpeted stairway, their arms linked together. Lady Denver is strategically touching her hair every once in a while, forcing everyone's gaze to her priceless necklace.

The couple is treading down almost to a beat, one step at a time. They halt majestically on the last-but-one step. Everyone showers them with flower petals. On the tread just below theirs, now stands Knight Henry.

He begins his address. 'Your Excellencies the Duke of Manchester, the Earl of Talbot, respected barons and baronets, our dear knights and all our ladies and gentlemen. As we celebrate this twenty-fifth marriage anniversary of Lord and Lady Denver, let's take a walk down a path filled with sweet memories from all those years ago. Two boats gently glide past each other on the River Meuse. One holds a strapping youth of twenty-five, the other, a tender blooming eighteen-year-old damsel. They are both with their families on these rides. For some unknown reason, the boats slow down when next to each other, face to face, while the boatmen converse for a few moments. And in that instant, in the glow of the setting sun, the two young people look upon each other, with the same innocent gaze, filled with wonder, just as Adam had gazed upon Eve. A few months later, twenty-five years to this day, this incomparable couple was joined together in holy matrimony with God as their witness.'

Lady Denver whispers to her husband, 'What a beautiful love story, who is he speaking of?'

Lord Denver mutters, without turning to look at her, 'Us.'

Lady Denver turns her head to look at him, weakly.

'Smile, everyone is looking at us,' he commands.

She smiles instantly, putting back her doll-like expression.

Knight Henry is droning on, 'In memory of that lovely day, the cake cutting ceremony will take place today in the middle of the Garden Lake. Our dearly beloved Lord and Lady Denver will be in a boat with all of you showering your blessings on them from around the lake. The first kiss after the cake cutting will bring the glorious experience of that first gaze alive again. We will all relive that day by showering the couple with flower petals.'

Lord Denver looks at Knight Henry with boundless appreciation. And Lady Denver says, as she digs the knife into the cake, 'When did it all happen, my love story with you on a boat ride? I'd first looked upon you through the hall window, as you sat in the garden. Your back was turned towards me. My mother had said, "There's no need to look any more, this is the man you will marry. Your grandfather's last wish." Although, I was sitting by my grandfather in his last moments. All he'd wished for then was to taste some vintage again. And he'd passed on even before the spoon was held to his lips. He'd said not a word about my marriage.'

'Ah ha, let that be. Isn't it a beautiful fairy tale nevertheless! I most certainly would have loved for this to have taken place. Even to this day, in truth, I see you as a glorious swan, gliding on these waters, a glittering diamond necklace around your slender neck.'

Lady Denver's cheeks take on a red redder than her lips. She says promptly, 'But now you must bring me pearls from the Indian Ocean. I've heard those pearls are as large as pigeon eggs.'

'No more, darling, those were gone hundreds of years ago.'

'I know they are still to be found, I've seen photographs. Be honest. You don't want me to have them.'

'Of course, I do want them for you. Who else will I gift them to? But today you must talk only of this fine diamond necklace, right?'

Flower petals rain down on the couple. And Matilda says to Ruth, 'I tore up the Sunday invite. Do get one for us, if you find it possible. I might join you if I can.'

YEAR 1788, OXFORDSHIRE

Surveyor Webb has seen the boy twice earlier. One day he was sitting under a cedar and writing accounts in a notebook, completely lost to the world. On another, he was seen sitting on a rock by the waterfall. On that second day too, he had a notebook in hand, although he wasn't writing much, just staring into the water. Is it poetry that he is writing? Quite possibly. Boys at this age do that often enough, particularly if in love. Of course, no girl has ever been seen with him. And why only girls, he isn't seen with anyone, not even boys his own age or even older. A lover of nature? Of open fields, gurgling rivers, burbling falls and deep silences? Does he not need people?

But even if a nature-lover, what does he jot down all the time? Has he even noticed the beauty around this morning? He's perched on top of a large white boulder. After a while he walks around and brings back several stones. He turns them around carefully, one at a time, and makes notes. What a strange boy! Sitting without a care under the blistering sun with not even a cover for his head. He doesn't seem bothered about getting drenched in case there is rain. Webb is astonished.

He's keen to speak with the boy but what should he say? Webb's headed to Marcuse's home, a wealthy farmer in the village. His assistant Aden is with him.

He could call the boy to check about the address, muses Webb.

'You there, boy, can you come here for a moment?'

The boy has heard him. He puts down his notebook on the stone and walks up to Webb.

'What are you called?'

'William.'

The boy doesn't want to give out his surname. Perhaps, he's from the lower classes. He's clad poorly, with a torn pocket and soiled collar.

Webb asks, 'Would you know where Marcuse lives?'

'Which one, Bob or Charles?'

'Charles Marcuse, the one with the purple grape gardens. Do you know him? You needn't come, just point us the way.'

The boy points to the north. And begins, 'Walk 1750 feet straight ahead. You'll come to a dark and bushy Alder tree. There's a dirt road by its side. Walk on for 740 feet and you'll come upon a red stony path. On your right, along the side, will be a tall mound of slate. On this path walk some more, around 2945 feet and then you'll hit a granary. There'll be roads on either side of this. The one on the left is slightly winding, around 1810 feet, and it has a waterfall along its limestone facade. The water has carved some amazing shapes along this, some like a horse's head, some like teacups, one even like a person kneeling in prayer.'

Webb's eyes widen in wonder. He's been to Charles Marcuse's home quite a few times, but it startles him now that he's never seen the road in quite this manner, with none of its details like how this boy has mapped it.

Hiding his surprise, he asks again, 'How many feet did you say we should start with?'

'1750 feet. And then a bushy Alder tree.'

'Wonderful! But would you have measured this distance by hand? Or is it just a guess?'

'No, of course not, not by hand. I've created an instrument. That is what I used.'

'Would you show me this instrument?'

'I'm not carrying it. It's at home. My folks are displeased if I carry it around. I sneak it out sometimes though.'

'I see. But you could explain it to me, right?'

'It's a box resting on top of a three-legged stand. The box has a hole through which I look at a distant point. I mark the distance, five feet at a time, with a string. And then again. One can shift the box up and down because the path is never flat. I shift the angle up as needed, down if that helps.'

'Have you ever seen a theodolite?'

'Only in a photo, yes.'

'What you have built sounds very much like a theodolite although that has a telescope attached to it.'

'I know, I've read about it in a book. About the first station from where we start the measurement, and then the benchmark, about baselines, bearings and so on. If only I could lay my hands on one, I'd measure up everything around.'

'If only you could lay your hands on one? Well, then! Which village are you from? What's your father's name?'

'My father's dead. I live with my uncle. He owns a field, and a dairy farm, too.'

'What's his name?'

'William Smith.'

'Cannot be. No one of that name has applied for their field to be mapped.'

'His field was measured some years back. Are you then the gentleman who has come to measure the village lands?'

'I am Edward Webb, Surveyor.'

The boy's face lights up instantly. 'You will be surveying Mr Hastings's father's estate, isn't that so?'

'So, you know that as well? Mr Hastings has indeed sent me a letter, but he is still in India. I cannot start the work till he himself returns to authorize me.'

The boy's face crumples somewhat. 'Will you teach me survey work, sir? You won't need to pay me a salary. I only wish to learn.'

'Yes, indeed, I could teach you. You seem earnest enough. I do require a person who knows the village well. How far have you studied?' queries Webb.

'I've studied with the village master, Billie.'

'But how far? You seem old enough to have completed high school. Which college do you attend now?'

'No college, sir, the village does not have one.'

'Is that so? So how far have you studied?'

'As far as I could. I did my utmost to put together money to buy books, and those I read.'

'Now listen, boy! What was your name again?'

'William. My uncle is William Sr., and I William Jr.'

'You seem smart enough, boy. When I ask how far, I want to know up to what class. Why is that so difficult to answer? Even if you have not attended high school, you must have attended some classes in the village, right?'

'That I do not know, sir. But I have been reading a book for a while, sir, Danielle's *The Art of Measuring*. I have purchased this book, it's mine, I do not need to return it to the shop. Should I show it to you?'

'No, no need for that. I read that book after finishing high school. I cannot figure out how you managed to read it. All I know now is that you possess no educational degree, but you have learnt somehow to read and write. And that is incredible. But I must be on my way now. Thank you for your directions.'

'You will not employ me then?'

'Dear boy, you need to be qualified to do survey work. We do need labour to help in the measuring and mapping, but we have already found those. The villagers have helped with that. It is better for you to enroll in school instead.'

'One needs money for that, sir.'

'Of course. There are some free schools in London but you will have to live there to attend those classes. That too costs money. I will proceed now, goodbye.'

Webb begins to walk away. He sees no need to converse further anymore.

A little while later the boy calls out again. 'Sir, please. Please, Mr Webb, do stop.'

The boy is panting as he runs towards Webb. What a pest, truly, he feels irked.

'What is it, William?'

'Please, sir, read this.'

'What is it?'

'My notebook. Where I jot down everything.'

'Whatever will I do with your daily accounts?'

'Not daily accounts, sir. I observe a lot of things, sir, and write down whatever I do not understand.'

‘What the deuce! What will I do with your questions about why the sky is blue and the grass green? You must study to know the answers to these.’

‘Not those, sir. Not about the reflection and scattering of light. Isaac Newton has written about it, many others after him too. I would not trouble you if there were answers to my questions in books or even with you.’

‘Alright, give me the notebook. Take it back tomorrow. If you come by eight in the morning, it’ll be fine. If not, I’ll leave it with the servant. Do you know where I live?’

‘I’ll find it, sir.’

‘What if you can’t find it, or forget?’

William chuckled. ‘This is the only notebook I possess. I scribble as tiny as I can. There are still two and a half blank pages left. It cost me two shillings. I won’t be able to collect that much money in a month. I will certainly find out where you live, sir!’

Webb puts the notebook in his leather satchel. It takes him ten minutes to get to Marcuse’s. Webb has rapid strides. He takes the shortcut on the right of the farm, a path he has taken before.

Marcuse is waiting in his garden, along with a few more farmers.

‘I apologise for the delay.’

Marcuse answers, ‘We sent men looking for you just in case you were lost.’

Webb chortles, 'A surveyor could lose his home, land, wealth, even his wife perhaps, but never his way. In truth, I chanced upon a boy from the village. I chatted with him for a while…but never mind. Tell me, has survey been done in this village before?'

'Well, nothing was mapped with a surveyor, no, but some government officials did visit after the Parliament Bill was passed. At that time, some villagers had smartly got their land measured and mapped. Most of the land on the eastern side of the village belongs to William Smith.'

'What was that name again, William Smith? Does he have a nephew?'

'Not one, but three, and a niece as well. His brother John's children,' replies farmer Timothy.

'Don't the children go to school?'

'Not just them, there are many here who don't. There's only one teacher, a man named Billie, in the village. The children are all sent to him. One odd child might end up in town, the rest continue here, tending to land and cows. What do we need to study for? Our sons wouldn't help on the farms if they became lettered,' Marcuse answers.

'But it must be said that William's oldest nephew had a good head for studies. The uncle's a penny-pinching skinflint. Does not even give them a square meal. The poor lad has laboured endlessly at my farm to save money for books.

I would give him bread when he finished. He would stuff it into his pocket and sprint to the bookshop. That devil of an uncle!' says farmer Udo crossly.

'Such gossipmongers, you people. It can take a whole day if we begin to criticise William. He's not a good sort, don't call trouble upon yourselves willfully,' Marcuse admonishes.

Suddenly, there is silence. Marcuse understands money after all, and time is money. They've all put in time and added the pounds, one after another, to bring the surveyor over. To waste that effort in chatter will be such a shame.

Webb pipes in immediately, 'So whose farm do we start with?'

Marcuse replies, 'Udo's land, in the northern corner. Take Mr Surveyor with you, Udo. After yours is mapped, he will work on George's. We must get both lands measured by this evening. Don't squander the good man's time with your foolish prattle.'

Udo nods, like a dutiful child. After all Marcuse is their leader. He has smartly managed to warn off the surveyor too—he will have to complete the work he has promised to.

Mr Webb works with an assistant, Aden. A fine worker, but a habitual drunkard. Aden walks up to Webb, reeking of booze. Webb is at a loss. This was the only man willing to accompany him to this sleepy village. Fledgling surveyors hate to leave London.

The work is complete by one that afternoon. Udo is delighted. Mr Webb hasn't looked up even once in the last three hours, so deep in work he has been. His assistant too has worked himself to the bone. Udo's parched just thinking of all the work the two men have put in.

Mr Webb is slumped under a tree, catching his breath. He has carried a ham sandwich with himself. He requests Udo, 'Would you have some basil at home?'

Udo steps inside. A little later his wife walks out with a tray in hand, her face suffused with a warm smile.

Webb asks, 'Could you find some basil?'

Udo laughs, 'You don't need to have that dry sandwich, sir. Have this warm lamb meat with bread, and some homemade wine.'

'Thank you, but no, Udo. I will not be able to work if I take such heavy food. Please, do take it back.'

'Gee, sir, this won't be on offer every day. At George's it's doubtful you'll be given even a cup of coffee. And our Marcuse, the stingy bugger, don't expect even warm water there. Didn't you notice how peeved he was when we called William miserly. They are, after all, birds of the same feather.'

'Never mind. They must make up for what they are spending.'

'So what? Should that stop them from treating men well? Isn't there respect for an educated man? I don't like such behaviour. Come now, have some food. Fresh lamb from our farm. Marinated in cider and roasted. Please try.'

'I will then, but just a little.'

'Please go ahead while I check on your fellow. Rest a while too.'

'Listen, Udo, Aden is a little too fond of his drink. Do not give him any wine.'

'Am I insane? I won't, not at all, now that you've said it. I'll just be back.'

Aden is sitting by the carrot patch, and he has no food with him. He's nibbling on a couple of baby carrots.

Udo calls out, 'Hiya, there. Don't spoil the patch. Come, relax and eat some lamb and bread.'

'Is it on you? If only there were more villagers like you.'

'More? None at all! I am the king of hearts. And when you are with me, there'll always be food. And how hard you worked! How does one repay that? I for one turn thirsty when I see someone toil so much.'

'Well, thirsty you may be but offer nothing to quench it with? Will lamb and bread relieve my parched throat?'

'No, mate, I can't share the wine with you. It's your boss's order. He says you love your booze.'

'Did he say that?'

'Yes, he did. No wine for you. How about ale though? Should I get a bottle of that?'

Aden is thrilled. 'Who the heck cares for wine? Bring on the ale. Wine's for gentlemen, and that we are not!'

'You've said it. A country yokel like me, and wine? It's the missus who wants to climb up in life. Let me run and grab that ale.'

'Will a bottle be enough?'

'Have no fear, first one, and then two! Just as your boss commanded, no wine, it's just ale.'

After they've downed two bottles, Udo now begins to feel anxious about Mr Webb. 'Oh dear! How unforgiveable, that a host should be drinking while my guest waits. What a shame!'

Aden slurs, 'A very grave error, my friend, while my master waits. Go, run, look after him.'

'And what about you?'

'What nonsense, am I not a guest, too? You are the host, you must go. You must look after his needs. Just bring me another bottle before you leave, or you will sin greatly.'

Udo rushes to Webb. He's quite fine even after the two bottles; there's only a slight throb in his head.

'Forgive me sir, I didn't check on you. Did you eat anything?'

'I'm fine, Udo, call Aden. Where's the chap? We need to head straight for George's now.'

'I'm afraid Aden's a bit out of action, sir, after three bottles of ale.'

'Oh no! Hadn't I warned you not to give him a drink?'

'Of course, I remember, sir. I haven't defied you. You said no wine, and I didn't give him wine. In fact, he too is like me, sir, he doesn't like wine. I gave him ale instead. Wasn't I right to do that, sir?'

'Very right you were! All the work's now gone to drain. What am I to do now?'

'Just rest here, sir. You need to do nothing. Just drink some more wine and take a nap.'

'Forget the wine now. I'll sit here and read the notebook. The boy was hopeful that I'd do it.'

'Who, sir?'

'William Sr.'s nephew, William Jr.'

'Our mad William?' guffaws Udo, and he just won't stop.

'What makes you laugh so?' questions Webb.

'The boy's a bit batty, sir. His uncle took him to work at the farm but all he cares about are rocks, and more rocks. He looked at the poundstone meant to measure butter and insisted that it's something from a dead sea animal, a foss or hoss, he calls it, I think.'

'Fossil?'

'Yes, yes, that's it. Now you tell me, sir, can a dead person turn into stone? This house has seen some dead people, my father, my mother, grandma, too, and gramp...no, he died twenty miles away...but did they turn into stones? Like statues?'

Udo transforms into a statue too, perhaps to enhance his point, with not a wiggle out of him.

Webb stands up and walks away. The path to the left of the farm is now on his right. As he walks along, he arrives again at the burbling spring on the limestone front. The water's slightly acidic. It's eaten into the limestone and there are gaping holes along the façade. And among the holes is the amazing exact shape William had mentioned, a person kneeling in prayer. Webb settles down under a large stone and begins to read the notebook.

June 19, 1784

Today I saw a pile of stones in Sarsden village. It's called 'Fool's Gold'. Not gold, but a form of iron blended with copper ore. Name: Pyrite.

December 12, 1784

Can't find hard rocks in Churchill to sharpen knifes. These are volcanic rocks. But there have been no volcanic eruptions here.

January 11, 1785, to January 19, 1785

I am finally successful. Village farmers spend money to paint their homes. I told them there's no need to waste money. Take blueish mud from the riverbanks and use as paint. Why is the mudstone so blue? Is it a mix? Who mixed it? What mineral is it that makes it so blue? Lapis? I have not beheld Lapis myself but seen photographs. It is surely mixed with something. I need to find out what.

April 3, 1785

A road's coming up near Henley on Chiltern Hills. I stand for hours watching the work. What is this that I see. Layer upon layer of rocks. White, red, ashen. What a beautiful sight! Stratified soil. The topmost layer is a dark brown. Then one after the other, colours change. Finally, a layer of rocks. Rocks crumble and become soil. What wondrous change is this that nothing remains of the stones? Even stones change like this. One stone becoming another. I love this. If only the sun shone all the time, I wouldn't sleep at night. I could watch this all the time.

July 21, 1785

Rivers, how beautiful they look. How disciplined their speed is! I could spend my entire life just watching it flow. Dan, my brother, says I share stories with the river. One could speak with rivers about joy and sadness. I could share all the pain in my life with it. My pain flows along with it and reaches the almighty. Rivers and winds reach man's prayers to God. But I don't recall my pain when I am with the river. I can find no words then. I just keep gazing at it. The river looks like something when seen from up close, different from afar, and still more changed from the top of a hill.

Let me explain. From up close, the river is like a flat tabletop. From afar it curves like our village streets. And from the hilltop, it looks like a wriggling snake. All rivers are not the same though. Some fan out like a palm, five fingers forking out from the wrist. Some are like vertical lines running through cracks. Or a straight line, and a hill on its side. I once saw a river, a full circle surrounded by mountains, its streams coursing in all directions, like gleaming rays.

I'd measured a river's width against a maple tree once. By drawing a grid using geometry. The river to the north-east of this village is black in colour. With a bed of black stones. Its sands wavy. Like wavy sandstones I've seen elsewhere. Did that have a river too? A river that coursed like this one? Who created these? What is the mystery behind creation? Has someone created this earth?

Pages and pages of such thoughts. Webb sighs. He walks back again to Udo's place. Udo's wife asks as soon as she spots him, 'Will they wake up soon? Just look at those two, snoring under the tree!'

'Forgive me, but do you have needle and thread?'

'Needle and thread? Of course. Do you need it at once?'

'Yes, please. But I will return it tomorrow. Before the evening. I don't know any thread shops in the village.'

'What are you saying! You don't need to look for a shop, I'll get it this instant.'

William Jr. is back home before sunset. His aunt won't allow them to use light until it turns pitch dark. He would write if he had his notebook, even in this twilight. He's feeling starved. But he can't ask for food, it'll drive his uncle mad. He's hardly done any work today. He can't go on like this much longer. Webb was right, he doesn't have a school certificate, he's never been to college, and will never do so. Who will employ such a person? An orphan, with three siblings. How can he make their lives better? There is no way, nothing to be done. Good riddance to that notebook. He needn't get it back at all.

'Brother, please put on the light, it's so dark,' pleads Elizabeth.

'Let it be today, Elizabeth.'

'Blimey, I don't like the dark.'

'What's wrong with it, it's nice. Get used to it. Will you be able to live under a tree?'

'Why should we live under a tree? Won't we have a house? Won't you work?'

'You, silly, who'll give me employ?'

'I hope my money pot is still there, behind Rosie's trough.'

'My silly pumpkin, of course it is.'

'Someone stole Dan's and John's. Wonder who. Poor things, they wept so much.'

'Who knows. There's a demon in the olive tree, perhaps he.'

'Don't, I'm scared.'

'He won't steal yours for sure. You're a good girl. Those two are wicked.'

'No, they are good too. Maybe a little naughty. Will you give them their shillings when you begin work? No need to take it from the demon at all.'

'I will.'

'Someone seems to be calling out for a William.'

'Must have come for uncle. Go, tell them to go to the main house.'

'Come with me.'

'I don't feel like. Go if you can, or else sit here quietly.'

Elizabeth goes out. She returns in a bit.

'Did you show him the way to uncle's?'

'Come outside, he's asking for you.'

'Who?' It's pitch dark all around. William cannot see the man's face. 'Who's this?'

'Brother, should I bring the light?'

'No,' William peers again at the stranger's face, 'the William you're looking for doesn't live here.'

A gentle voice floats his way, 'William Smith, you are henceforth employed under Surveyor Webb. You will be paid one shilling a day. Tomorrow morning come to George's land at the north of the village. You must reach by eight. Do not be late.'

The man walks away. William is at his wit's end. What was all this madness the man spoke of? Who was he?

Elizabeth is dancing a wild dance of joy. Elizabeth is Liz, and William Willy. But only Liz calls him Willy, and only when she is delighted beyond belief. Like now. 'What fun, what fun, a shilling a day, Willy! Who was that man, the demon from the olive tree?'

William stutters, 'Maybe.'

Webb is walking on, silently. He will hand over the needle and thread to his wife. She sews very fine. He'll give his linen shirt to her, a gift from his sister last Christmas. If his wife tightens it a little, an inch all around, it'll surely fit the boy.

YEAR 1788, HASTINGS ESTATE

The village is facing a rather massive crisis. Everyone is gathered at Marcuse's farm. If they can't bring the matter to a close, they'll call the police.

George is now shouting at the top of his voice. 'Is this a joke? I've earned every penny with the sweat of my brow. And to squander it away like this? And what rubbish measurements are these?'

Mark adds, 'Who are we paying for? We had to pay more for a famous surveyor. And the man simply handed over all the work to his assistant. How could he add my pond to George's land?'

'And what about how he added my blood-orange garden to Harry's land?' George rants.

'No, no, we cannot accept this. We cannot allow just about anyone to map our lands,' this is Mark again.

'Ohh ho, what is there to be so confused about? We all know where our lands are...'

Mark flares up even before Udo can complete his sentence, 'Shut your trap. Stay quiet. It's all because of you. You got your land measured by Webb, so it worked out well for you. And then you went ahead and spoilt it for us all.'

'Now, now. There's no point creating trouble amongst us. Everybody calm down,' Marcuse tries to pacify the crowd.

'How can you ask us to be calm? We agreed to get our lands measured only because of you. An hourly rate for the surveyor was fixed. And we paid up. But there's been no clue to his whereabouts for the last two months. After working on a few properties he handed the work over to an assistant. And the chap knows nothing. He's making a mess of all the measurements. A half bovate land of mine has become five acres.[1] A whole half missing.' This was Harry's lament.

'David owns only a rood of land, and that's become two acres after the measurements. And David is dancing with glee.' This was Mark's complaint.

'Why has he taken another employ when ours is incomplete?' George adds.

Now Mark demands again, 'Has he really taken another employ?'

---

1 A bovate is an old English unit of land measurement that is equivalent to one-eighth of a ploughland or carucate. The size of a bovate can vary from 10 to 18 acres, depending on the condition of the land.

'Don't you know? He's at the Hastings's Estate in Daylesford, Worcestershire for the last couple of months. 650 acres of land is being mapped.'

'Why? Is Mr Penistone planning to sell his estate?'

'That, of course. But to his son. I've heard the son has made pots of money abroad, all through dubious sources,' George winks.

'What's all this? It's not fine to bad-mouth a respectable man,' interrupts Marcuse.

'Keep your goody-ness to yourself. I'm saying what I've heard. He'll be thrown into prison as soon as he steps on English soil, y'know? There are many complaints against him.' This is George again.

'We haven't gathered here to slam the Hastings. Let's talk work.' Marcuse observes, his voice severe.

'Alright, just as well. Where is Mr Webb?' questions Mark.

Marcuse replies, 'He's been asked to come back, he'll be here shortly.'

'He's meant to be here in the village. Let's all go to his house, should we?' George retorts.

'We all know that he's not here in the village. He will come back for sure.'

Just then a small boy comes sprinting towards Mark, panting, 'The man's coming, Papa, I saw him near the spring.'

A little later Webb arrives, his face a ripe red from the long walk under the hot sun. He sports a two-day old stubble. He folds his hands instantly, 'Please forgive me. I've been at fault.'

'What do you mean by that? You're charging us for work and then you hand over to an incompetent nincompoop?'

'He's extremely competent, I promise. I can't figure out how he could make such a grave mistake.' Webb's voice drips with apology.

'That matters not at all now. Either you return your fee or complete the work pronto.'

'I'm afraid I will not be able to return my fee. My assistant ran away with your money last evening. But I promise, I'll complete the work myself.'

'Fine, you have a week for it.'

'That'll not be possible, a week is not enough for a job that'd take at least two months. And I've committed elsewhere in the meanwhile. I promise to start in three months, please.'

'We cannot allow you to work anywhere else, Mr Webb. It's not possible for us to wait another five months to have these maps ready.' Marcuse seems grave, 'You must finish up in a week. You cannot step out of this village till then.'

'Please believe me, that's not possible. If you force me, I'll call the soldiers,' Webb speaks with unbelievable calm.

'Call them. We will hand you over to them ourselves,' Mark threatens.

'Why are we waiting? We still live under the King. Let's punish him like we did a few hundreds of years ago,' George butts in, 'hang him by the ropes and beat him with hot iron rods.'

No one discourages George, in fact, even Marcuse falls silent.

Edward Webb touches his right hand to his chest, closes his eyes and sighs, 'Dear Lord!'

Someone raises his hand, and a soft voice floats over the commotion, 'I want to say something.'

'Who's that? Come forward.'

The boy comes forward, working his way through the crowd. His face is shrunken, burnt red from the sun. A dark stain is visible on the tip of his nose.

Marcuse barks, 'You, lad, what is it?'

'Pffft. What can he say, drifting around the forest all day, the bugger!' George is his acerbic best.

Mark interrupts, 'At least give him a chance. Speak, boy, aren't you stingy Willie's nephew?'

'Crap, yeah?' Marcuse butts in. 'You're bent on having the village knock each other out.'

'Don't behave like a priest, Marky, you've no problem in calling the blind and lame blind and lame. But not the stingy?'

Now Udo offers his two bits, 'Why don't you give us coffee, Marky, we've been here so long. It'll never be ale, we know.'

Marcuse's eyes could burn holes into Udo, then he walks away home.

'Now look what you've done. Why do you speak nonsense? Let's discuss the real problem which has nothing to do with you,' George cuts through.

Udo too walks off in a huff.

'I'll survey the land, within seven days. Please let go of Mr Webb,' the boy says.

'You, you half-pint? What do you know of surveying land?'

'I know enough. I've learnt the work in the last two months. Ask Sir if you don't believe me.'

'Sir? Of course he'll say yes. He'll scoot after he deposits you here.'

'I will never agree. It's not humanly possible to survey 600 acres of land in a week.' Webb seems immobile to William's claim. 'I'm very grateful that you've come forward, William, but please do not say yes to this. You don't have any idea what you're agreeing to.'

'I'm saying yes to what I can do, Mr Webb. This is my village. I've grown up here, I'd be happy to do something for these people. And you are my tutor, it'll hurt me if I bring disgrace to you.'

'I can't find words enough to bless you, my boy. May God bless you. But I'm afraid you'll fail,' Webb's voice quivers.

'What's with all this drama? One says can and the other can't. And both are trying to make fools of us,' George is livid.

'Give me those maps you're holding. I'd like to get some work done before sundown.'

'You'll get the maps, but we won't let your master off. He must remain here for seven days. We'll not pay a penny for this,' Mark repeats.

'I'm not leaving. If I must starve these seven days, I will, but I'm not budging. I'd like to see the end of this, too,' Webb says as he stares at William.

'Why would you starve yourself while Udo is alive, sir? You will dine at mine both meals,' Udo is back, a large clay jug in hand. 'Here's a glass of palm wine for everyone. The sun's scorching today. Where's that miser Marky, gone across the oceans to get coffee, yeah? Call him.'

Marcuse has heard all the noise, and now he peeps out. He makes a sour face and says, 'Why did you have to get wine, Udo, I've put the pot to boil.'

'The coal must be wet if it's taking you an hour to light the stove, hey Marky?'

Marcuse nods, his face pitiful.

Webb makes a request, 'Please assure me, Marcuse, that if a messenger comes from Hastings Estate, he will not be driven away.'

The meeting breaks up. William is headed to the north of the village, to Joseph's farm. Udo runs after him, 'Come and have a bite in the evening, boy, if you can't now.'

William Smith gives him a smile. 'Thank you, uncle, but I'm used to going hungry for two or three days at a stretch.'

'You may not need it, but I must insist. Had I not tempted Aden with ale, we would not be in such trouble. It's all my fault.'

'Don't be sad, uncle, Aden was a drunkard anyway. You're a big-hearted man, so you fed him. How many are there like you?'

Udo's eyes brim with tears. 'You mad boy, come by this evening for a hearty meal. You'll fall seriously ill if you work on an empty stomach.'

The nine-day long captivity in the village ended last night. Hastings Estate is still nearly thirteen miles away by foot. Webb and William are walking towards it briskly. Mrs Webb and her pet are trotting along in a horse drawn cart. The villagers have paid for it. Rather, paid very happily for it. William is sporting a feather cap. A gift from George. A large supper was held last evening in a village sozzled on ale, dancing wildly to the beat of music. It was almost like Christmas night. Even Mr Webb had joined in. And Udo? He had transformed into a mad, mad man, prancing with William on his back. And the boy himself, surrounded by the madness being celebrated in his honour, had been sitting shyly in a corner somewhere. Even the miserly Marcuse has gifted him a new notebook. William is yet to open it. Right now, he's yearning to meet Penistone Hastings's son.

'What does Mr Warren Hastings look like Mr Webb,' he asks Webb.

'I haven't seen him myself. But Mr Penistone was a poor farm hand in your village, in Churchill.'

'Oh! Then a farm hand can become a king, too, right, sir?'

'King?'

'Why, like Warren Hastings? Not a king here, but of a country, nevertheless. Everything happens because his word is law.'

'No, not everything. India has many such kings. But, yes, he did have some powers. He is indeed a self-made man. He's educated, and a polyglot. How does lineage matter?'

'It matters, it does, you know that too Mr Webb. Never mind. How much is the Estate going for?'

'Eleven thousand pounds.'

'Oh.'

William walks on silently now. There's much he cannot express, he may never have had the chance to do so.

Webb says again, 'It seems Mr Warren Hastings is returning to England with eighty thousand pounds. Could be a rumour. But it is confirmed that he's taking over the estate for eleven thousand pounds.'

William is silent.

'I've heard he's waged several wars in India. He sends soldiers to seize properties from local rulers. He's punished them unjustly. But those who are impeaching him now, are his political rivals.'

'Are there big and small dynasties in India, Mr Webb?'

'I wouldn't know in detail but when there are rulers, there are also the ruled. With the powerful exist the powerless. How else would one rule? Mr Hastings had joined as a simple clerk in the East India Company. But he's brave and enterprising, qualities not many have.'

'Has he called us only to survey his land?'

'Surveying is needed, of course, but that's been done already. He wishes us to design a landscape for his estate. Do you have any idea about landscaping?'

William shakes his head.

'I'm very happy to see that you admit to what you don't know, William. Always own up to your mistakes. After that make amends. I have a few landscape sketches of well-known spots in London and Paris. Study them carefully. 650 acres is not too large a land. I've worked with land as large as 2000 acres. One can build houses, palaces, forts. We must understand what Mr Hastings requires. Does he need an office, where would he like the staff living quarters, the stables, would he like an enclosure for other animals.

We must know each detail. I've collected a few sketches of Indian palaces. What if he asks for an Indian-styled home? What would the best spot for this be? We must have everything decided before making the sketch. Mr Hastings loves horses. We should check if he wants a racecourse or not. Although 650 acres is not large enough for a racecourse. We have the next seven days to discuss all these details with him. In that time, you should study more about landscapes. You could discuss it with my missus too. She's helped me in the past with designing.'

'When will we meet Mr Hastings?'

'Tomorrow, but I'll be going alone. You should not meet him till you have a deeper understanding of the subject. Mr Hastings is not staying at the new estate. I'll meet him at his old residence. They've put up two tents for us in the estate compound. We will stay there.'

***

At last the day dawned. For the last six weeks William and the others have spent their days in the temporary tents inside Hastings Estate. William in one, and the Webbs in the other. Warren Hastings is a man conscious of rank. William's tent has a small room, with a tiny space behind it. The Webbs are in a much larger one. The last few days, Webb has been going off somewhere after the noon meal. When he returns, it is already dark.

Last evening he had come to the tent and said, 'Have you completed the landscaping of the fort, William?'

William put the drawing on the table. Webb did not look at it. His face dripped with tension. 'Have you consulted Mrs Webb?'

'Yes, she's made a major change. The dining room is now to the left of the ballroom. Now, the dining room will have a full view of the lily pond. If we know how many varieties of water lilies we need, it'll aid the landscaping of the pool.'

Webb seems distracted. Perhaps, he's heard nothing of what William has said. Mrs Webb's dinner bell can be heard now. Webb goes and sits silently at the table, his arms placed squarely on it. He'd shut his eyes during the supper prayer. But now he shows no signs of picking up his fork and knife.

Mrs Webb questions him casually as she ladles soup into a bowl, 'Have you lost the Hastings Estate project?'

Webb's suddenly alert, smiles lightly, and responds, 'Not yet, but may soon.'

'It seems there's a warrant out for Mr Hastings.'

'Not thus far, but it's expected.'

William continues to eat mutely. He's extremely eager to know more, particularly about this man from Churchill. A farm-born man, risen so high! What crime has he committed in India, for which he is being hounded in and out of the court of England? What offence is it that has called for impeachment?

'The newspapers say that he was felicitated after his return home,' points out Mrs Webb. 'He's received high praise from the board of directors of the East India Company, indeed, from the King himself.'

Webb counters, 'But Sir Philip Francis of the Calcutta Council has made several charges against him.'

'What is the Calcutta Council?'

'A council set up in the country where he was serving. Edmund Burke will prosecute him.'

'Philip Francis, Edmund Burke, names I've never heard of. Who are they?'

'They are members of the East India Council Board. We may not have heard of them, but they are extremely powerful citizens.'

'Why should he be punished for what he has done in another country?'

'Because he was Governor General there and has now resigned from that post. He unleashed many wars in that country, and enormous amounts of Company money was spent on those.'

'Why so many wars? Were there many rulers fighting?'

'Yes, uncountable numbers of rulers. Maybe close to two hundred of all kinds.'

'Is that even possible? What is this country India that you always speak of?'

'You could call it an administrative or geographical area.'

'A country?'

'Yes, that.'

'Sorry, Edward, but how can there be so many kings if it is one country?'

'I too do not understand fully. Perhaps their kings are unlike ours. The Company has seized some kingdoms, on some others it exercises indirect rule, and yet some more remain independent.'

'Dear God, why would the Company even go to such a strange place?'

'It's all about trade and business. They simply loot untold riches, or pay for it cheap, and transport it here.'

'Well, let them do business, why get into war?'

'It's bound to happen. While one king gives an opportunity for trade easily, another is too stubborn to do so. The difficult ones need to be subjugated.'

'Why would anyone go all that distance to set up government? Why take over the entire country? In any case, if you wish to trade, do that, and get back. Does any of this sound logical to you, Edward? If you pluck fruits from my garden and do not pay me a fair price for it, that's not trade. It's loot.'

'Look here, I do not know too much about the East India Company, not what they do, how they do it, and why just me, perhaps ninety per cent of the common people would not be aware of this. We are discussing the Company only because of Mr Hastings's case. To speak the truth, I'm not at all curious about them.'

'Fine, then, let's discuss what arouses your curiosity. The matter of how our country is getting inextricably involved with the civil war in America. Why would it be so?'

Webb is done with his food in an instant. Wiping up with his napkin, he says gruffly, 'We are here to work on the landscapes for Mr Hastings's estate, not to discuss the intricacies of politics. I must be at his office early tomorrow morning. William will accompany me. I'm very tired. I'm afraid, I cannot carry on this conversation further. Good night.'

Mrs Webb continues with her food, her face heavy. Edward walks to their sleeping quarters. Mrs Webb turns to William. He's staring into his food. He has no desire to meet her eye. His insides are all aflutter. Tomorrow morning he will finally come face to face with the man he has been waiting so eagerly for. Will he be atop his horse, his sword drawn? Is he a man easily angered? Will he strike at the slightest provocation? Or pull a gun?

He spills some soup.

Mrs Webb speaks up, 'Perhaps you are full, William. You should retire for the night.'

Like a good boy William stands up.

'I've stitched a shirt for you, and a leather cap. Wear them before you leave for Hastings House.'

William mutters a quick yea and rushes out of the tent. Outside it's sparkling with moonlight. Is it a full moon tonight? It is on nights such as this that the Lunar Society members come together in Soho House. What if he could join them? But he has nothing that qualifies him to do so. Even Mr Webb has never been invited to join them. What if they do invite William one day? Will there ever be such a day? Will William ever rise, beyond what Mr Hastings has achieved? Not with wealth but with his knowledge, such that Mr Erasmus Darwin will himself pen him a letter of invite? William is like the clouds that are swimming around the iridescent moon. No matter how hard he tries, he may never enter its kingdom. A chilly draft fans his face, and he shivers uncontrollably.

His being is awash with discontent, one that rises from an unattainable, unsharable dream. No one caresses him, there are no reassurances, no one to say yes, yes, this too will come true. Why does he dream then?

Christ too was born low. But he was not motherless. William has no one. Perhaps the Lord deserts the orphan as well.

YEAR 1793, STOWEY

'Are you missing home, William?'

Webb has been forced to stop the Hastings Estate work midway to travel to Stowey. He has a project at Baronet Jones's estate too, in addition to some other odd jobs that need attention. Despite spending money like water, the erstwhile Governor of Hindustan, Sir Warren Hastings, has not managed to avert his impeachment. His case commenced on 13th February 1788. It's been five years since then. This afternoon, William has brought Lady Jones's work almost to a close and returned. He now lives with Webb. These days Webb does not ask him to accompany him everywhere. He has given William permission to carry on independently for smaller projects. In fact, William has pulled off the Jones contract almost singlehandedly, although the deal was made with Mr Webb's name on it and the entire fee was, therefore, handed over to him.

William looks a tad tired and dejected today. He's twirling his leather cap on his finger without pause. Immediately on his return, Mrs Webb has cleverly drawn him away to her rose garden on the pretext to show off a new cutting. He seems at a loss of breath after rattling off the last month-and-a-half's daily dossier. There is no change now in his expression at Mr Webb's question.

'I had no home to call my own, only a patch to come back to at night.'

'Do you remember your brothers and sister?'

'Yes, indeed. My sister very much. Do you know what the tiny girl had said to me? "Leave, brother. We will grow even without you. But you will not, never ever, if you remain here."'

'Do you wish to visit Churchill for a few days? Whyever not? Go, meet them all.'

'I wish to keep walking, sir, only that, and I yearn to become a map myself.'

'Become a map? What can that mean?'

'No, nothing. But look south-east at that barren mountain. There's a valley along its downhill slope. Look at the sunlight as it hits the mountain. The white rocks break into colour as the sun lights upon them. Midnight blue in places, sometimes lilac, and then at times a dark snuff. This year has seen many cedars bloom. But they've dropped all their cones because of unseasonal rains. The cones have rolled down the slope, southward, and are piled one on top of the other. If the slope of the rocks and the slope of the hill are the same...'

Webb's lost patience by now. He can't wait for William to complete. He cuts in, 'Of course, they'll slope in the same direction.'

'Not always, sir. Mountains are a geographical space. They peak and then descend all around. But look at the rock strata sloping eastward. A bed of white rocks overlying a bed of red rocks, both inclining east. Does it make sense now, sir?'

'Not too much, William. I'm a surveyor. Measurements are what I comprehend.'

'I too don't understand all of it. Something is happening or has been taking place for centuries. But its truth, its root, eludes me.'

'Who is causing it, William? Do you think it is God?'

William laughs, 'Causing it? I do not care for that. My question is, "why"?'

'Did you visit the estate of Baronet Jones's widow Lady Jones?'

William nods.

'I'm sorry, I didn't pay you the carriage fare. How much did you spend?'

'Thank you, sir, but that's fine. I walked all the way and back.'

'Walked! What route?'

'I took the Trans-Cotswold, then the Roman roads of Akeman Street, where I looked at some rocks. On to Cirencester through Burford, and southward to Fosse Way,

Tetabury, and onward to Bath along the old Norman colony. Then on to Rodstock, Siden, Stone Easton, Temple Cloud, and then the Jones Estate. All these villages have coal mines.'

'Unbelievable, William! Trans-Cotswold itself is fifty miles or so, the most unpassable path in all of England. And you walked it?'

'I am used to it, Mr Webb,' William gives a shy smile.

Webb laughs out loud at something remembered. William looks at him surprised. He's not one to ask many questions.

Webb offers the answer himself, 'I've reviewed your diary. Tell me now, were you disappointed when you met Mr Hastings?'

William snorts. He's embarrassed. 'Not exactly, I mean, he was astride a horse. Behind him sat a lackey, almost double his weight. The poor horse, a rheumatic one at that, could barely move. I couldn't quite figure out why he needed a lackey with him. It is beyond me why a war veteran would even need one. Strange indeed.'

'It seems he asked William to sketch him,' Mrs Webb now joins them. She may have heard their last bit of conversation.

'Well, if he's spending as much as he is, he wants to be sure of what he's getting, I suppose.' This is Webb.

'Doesn't know anything about maps, does he?' Mrs Webb questions.

'Who's said that to you? William? He's not one to speak ill of others, hey William?'

'Why would he speak ill of anyone? I had to prod and push and dig it out of him. Apparently, Hastings has asked William to increase the scale of the castle and decrease that of the pond. Where have you seen maps with so many scales?'

'He wanted the castle scaled up because he couldn't grasp it too well on a small map. A fat customer he was. Unfortunate that the estate work had to be brought to a stop. Heaven knows when it'll begin again.'

'Why so? The impeachment will end one day. And if he wins, he'll want the work restarted.'

'Winning might not solve his problems. He's spent tons of money on the case, Duke Ellenborough being his lawyer. Hastings may not be able to recover his losses fully.'

'Yes, entirely probable. How indeed does a man collect as much wealth as he did after only twelve years abroad! And that in an impoverished country like India. I've heard that the man on the street there has no clothes on his back.'

'That, yes, but its kings and emperors roll in wealth. Their palaces, forts, apparel all gilded in gold. If all their riches are put together, they can buy England twice over.'

'Has the man then plundered these kingdoms?'

'There you go again, telling tales. Let it first be determined whether the man is corrupt or not. I don't like to indulge in false accusations.'

'Why would you, after all he's a client.'

'Now, banish these thoughts. Cook something delicious for dinner tonight. Tell me what to get from the market.'

'Have you invited someone over?'

'No, no, it is to celebrate William. He won't be with us after tonight.'

'I'm baffled. Where will William go? Where are you running off to, William?'

William shakes his head silently. He's flummoxed too.

Webb says theatrically, 'Oh dear! I completely forgot to tell you both. Look how William put my mind in a tizzy trying to tell me about the stones and the rocks in the mountain across. As a matter of fact, we have received a new project. William's now been fixed to work in Lady Jones's mine in High Littleton. An independent salary. And it's work that he loves. He'll be able to examine whatever lies under the earth. I've already said yes without checking with William. You would like it, wouldn't you?'

William nods his consent.

Mrs Webb suddenly sprints toward the kitchen with a hurried 'Back in a minute.'

'Wait here, I'll go get the list. How about your favourite apple cupcakes tonight? Let me tell the wife.'

Mrs Webb is rinsing the cups and dishes. Edward leans against the dining table behind her.

'Don't be sad, darling, wouldn't you wish the boy to be independent?'

'Why not? Have I prevented him in any way, in fact, can I even?'

'Then why so miserable? Your voice is choking.'

'How does it matter to you both? Neither of you care a whit.'

'The boy has his entire life to live. We wouldn't stop him had he been our son too.'

'Why are you saying stop, prevent, and such? He is helping in your business. And you too need an honest assistant like him.'

'True, I need one. But that man is not William.'

'Why? You're the one to praise his work at all times.'

'He is excellent, truly. But William is not simply a surveyor. The boy's mind and thinking are unique. He will not be content working with me for long.'

'So, you understand what makes him happy, is it?'

Edward sighs deeply, 'No, darling, I do not know what will make him happy. But he is searching for something. I do not know what it is, perhaps neither does he. But this I do know, he can never begin his quest if he stays with me. He must be set free. I wish to do this for him.'

'What'll he do with his search? He has no one in this world. What if he is lost? What if he finds nothing in the end?'

'That's his destiny, unforeseen, unpredicted. But I cannot hold him back for that. Do you not see how restless he feels?'

'It's your madness and make-belief. I haven't seen another as calm as him.'

'A youth of twenty-two, maybe twenty-three, he stays up silently night after night. No beloved, no friends at all, distracted and withdrawn, something runs through his mind all the while. He looks on helplessly when questioned, as if woken from a deep slumber, forced to look alert. I would think you'd notice.'

'Why wouldn't I notice? A sad child, with no father, a mother as good as gone, no news from his brothers or sister. It hurts him, clearly.'

'Your affection blinds you. But something roils within him, churning so forcefully at times that he knows not what to do, how to put words to it. The feeling cleaves through him, its tremors will push him in his search. He is already questing, but I can't be of any help.'

'Why not? You are his teacher, his guiding light.'

'I know I cannot. But why not, that I do not know. And that is why I need to set him free. Indeed, I can set him free because I am his guiding light. Remember, too, that should we never see him again, do not despair. He does not fathom social responsibilities. He will return if he wishes to, if not, then this is our last time together.'

Mrs Webb smiles through her tears, her voice a whisper, 'You are as cruel as he is.'

'He is not cruel, no, darling. He's never been loved, or shown affection, he knows nothing about the difference between tenderness and cruelty. I can't say what he will do, except do not hope, do not wait, just release him. Or else it'll tear you up. I request you to not take promises from him.' Webb pauses, and then says, softly, 'Tell me, now, darling, what would you like to cook for the boy?'

YEAR 1794, SOMERSET

The village has grown one tiny bit at a time across fifty square miles of Somerset County. On its west lies Clayton, enclosed all round with red-brick houses, and to the east, Combe Hay. There's Griston to its north and Kilmersdon in the south. The village looks shabby, bare and depleted. The wind knocks against the mountain walls as it moves forward but fails to move a leaf. There's hardly a tree worth its name here. Had it been industrial, the wealthy would have opened their coffers to lift its fortunes like Lancashire, whose peaty soil boasts row upon row of apple and cedar trees. Typically, an industrial town hardly attracts poets, litterateurs, or even philosophers. But Lancashire is different. Many kinds of people have put down roots here. And everyone craves civic conveniences and comforts. But the run to develop Lancashire has left many other places miserably worn-out, and of them particularly is this village of High Littleton in Somerset.

The land is dug up everywhere. Deep pits and high piles of soil. Not a single landowner in the village is a farmer. They have come in droves only to dig. The land is pitted dangerously all over. If someone slips in, there's no saying how deep they will go. Some shafts spew poisonous gas all the while. The gases catch flame on their own, and burn, and keep burning almost as if someone has lit a fire but forgotten to put a frying pan on it. Some holes are gurgling with hot water. The land is scorched, its green singed with the heat and toxins.

High Littleton has no councils, neither at the village level nor for the city. The villagers hardly care for their own comforts. Indeed, why should they even be bothered? After all they don't live here permanently. They are labour, come to work in the mines here. Nothing matters, nobody cares, about trees, wheat, or even about grapes, or apples, or currants, maize, even the grass around. They work below the surface of the earth, subterranean beings digging their days away in darkness.

High Littleton is not an old village born of the bonds of landowners and cultivators. It is a village of mine owners and labourers. A place of loot and plunder. Apart from mangy dogs no living creature can be spotted around the huts of the workers. The only purpose ducks, hens, pigs or sheep serve is as food on plates or roasted red on skewers.

High Littleton is home to nobody. Almost eighty per cent of the village belongs to Baronet John Jones, who passed on some years back. His widow, Lady Jones, is the current owner of this property. The Jones estates are spread across many such lands including some in Stowey and Ramsbury. Lady Jones has lost her husband but not her courage. She runs each family business adroitly since she has been alone.

As the Director of the High Littleton Coal Company, she has engaged the services of a young man to carry out the work of survey planning and draining in her Mearns Pit mines. One usually employs experienced hands for such work. But Lady Jones has chosen an absolute novice with only a couple of years of survey work behind him. Her decision has been driven by the fact that most of her mines in the area have already been scoured clean over the last thirty-five years or so. More layers of coal can now only be found much deeper under the earth. But the cost of unearthing coal from that depth is exorbitant. The mines may even cave in any day now. And if this were to happen in the morning hours, many workers may lose their lives. She has, in fact, already brought in a physician from Lancashire for first aid in the case of such emergencies. The rest is fate. Lady Jones does not wish to bother herself unduly with High Littleton mines anymore.

These mines had already disgorged much of their profits while Baronet Jones was alive. There now seems to be no logic in engaging an over-salaried surveyor for these nearly hollowed empty pits.

The previous surveyor has been shown the door, and this new one has been brought in for half the cost. Lady Jones has been forced to get a new man only because of the workers. The absence of one forewarns them that the mine is soon to be abandoned. And they would be off overnight, ferreting around for jobs in other locations. If they band together, it'll mean big trouble. Every little bit that comes out of these holes is yield till the walls cave in. Somerset has been a fertile ground for long. It has provided England coal for several decades now. Many new mines of course are now being uncovered. In fact, entire England seems perched on an enormous mound of coal.

Most believe England will never lack for anything. Because its soil gleams with coal, its waters teem with fish. There's truth in this. People of the land have gone insane, with only coal on their minds. The boorish middleclass has begun to buy miniscule plots of land hoping to mine these. Rivalling for land in such numbers will be disastrous. It may have been a fair fight if it was between people of the same class. But this! The big owners have appealed to their King. Some months back a meeting was called at the White Hart Hotel to decide which mines would benefit, and which not. Duke Bridgewater had lorded over the conclave. Lady Jones can't stand the sight of him. She attributes the impending decline of Somerset to this man, this man who owns mines in Lancashire and has powered his way to get the Parliament to approve his plan to dig a forty-two-mile-long tunnel in his county. This tunnel transports coal from his mines to Manchester and onwards to the riverbanks of Marseilles. Now no longer does he need the London middlemen to get English coal straight to Paris. With that one tunnel he's brought the prices down by half. For untold years Somerset mine owners have had the better of the Bristol coal market, and that grip too now the wretched man has snatched away. Production in Somerset is close to shutting down.

Lady Jones gives the new surveyor a hard look up and down. He doesn't appear too shrewd. Serious but sincere. Lady Jones scrutinizes him as an employer would. The boy won't be too demanding. Her information is solid too. He doesn't boast of a lineage, and is an orphan to boot. He's unschooled, most of his knowledge derived from what little work he's done here and there. She needed one just like him. The qualified-certified ones are always asking for something or the other even as they provoke the workers. They even threaten to dump work. This guileless simpleton seems different. Even if he were to lose his life in a mine collapse, there would be no one to ask after him.

Lady Jones says, 'Have you worked in mines before, Mr Smith?'

'I haven't had the fortune, Lady Jones.'

'It might prove quite a problem then don't you think? I hadn't been told of this. I'd thought you to be experienced. You've done quite a wonderful job on the Stowey Estate.'

'It won't be difficult for me to learn quickly.'

'One cannot learn old mine surveys swiftly, Mr Smith. In truth, there's no one here who can teach you. If there's anything to learn, it'll be from the workers.'

'That's no problem, Lady Jones.'

'Hmm…well, you see, I will not be able to afford a surveyor's full salary for you, not until you have learnt well. You are untrained after all. You haven't a degree either.'

'I've worked with Mr Edward Webb for four years, ma'am.'

'Pffft! I'm aware of all that. But there's a huge difference, Mr Smith, between surveying land and mapping mines. Anyway, I cannot assign you surveyor quarters right now. You must live in a house east of Rugbourne village within High Littleton.'

'Thank you.'

'Now listen to me carefully. That house is not mine. It belonged to a Major Britton. The man went bankrupt on coal. You must well be aware that it's not a game to run a coal mine. If digging a hole in the ground were enough, there would then be no difference between a farmer and a baron.'

Lady Jones looks obliquely at Smith. There's no change in the boy's expression. It's a good sign. Best to tame the creature right away.

'In any case, Major Britton has disappeared, leaving everything behind. Harris, a farmer, took over his house. Although all the land around that area is now mine, I haven't managed to throw Harris out. I despise engaging with the lower classes on matters of property. However, if Harris agrees, you could stay there. The company will take care of your food. It'll pay the rent as also for your daily needs. I wouldn't know how much, but they'll pay nevertheless. Do you agree, Mr Smith?'

William Smith nods, silently.

'That's wonderful! Let me mention that the house leads to three different cities. A little distance away no doubt. Bath and Bristol are ten miles away, and Wales, around twelve miles. You'll have to arrange for horses if you wish to visit them. Which, of course, the company will not pay for.'

'That sounds fine, Lady Jones.'

'Perfect!' Lady Jones smiles, finally. 'In that case, you can begin tomorrow. I'll inform the store. Please take your ration for the next four days. But you must carry your own load, I'm afraid. I cannot give you a horse, or a carriage. Rugbourne is five miles from here.'

'Goodnight, Lady Jones. I'll see you in the morning tomorrow.'

'Indeed, Mr Smith. Welcome to the High Littleton Coal Company. It's four in the evening now. If you do not reach Rugbourne by six, Harris would have shut his doors. Do not delay, Mr Smith. Good night.'

YEAR 1819, LONDON

It's been raining in the evenings for the last four days around this same time, continuing well into the night. It is such a bother to wear a heavy raincoat and walk over the puddles, rainwater sploshing around. But today, John Cary is sitting in his office room waiting for the rains to begin. Curiously, it is not raining. The atmosphere seems heavy with all the signs, and yet, it is not raining. Cary's getting very restless. This was not how he'd expected it. Like most Londoners he dislikes endless rains. But if the rains evade him today, he'll feel very vexed because four of his printing staff have used it as an excuse and already taken leave from work. John too has promised at home to leave as soon as it begins pouring. Yet the showers have refused to come so far. He could, of course, leave now. He doesn't need to answer to anyone as he owns this press. But that's the real problem. He's snowed under with work these days and it doesn't seem right to leave with it all waiting. If he leaves now while it is still dry, what will he do when indeed it begins to pour indefinitely? And today, as he sits here without his workers, it's a drier day than usual! How impossible it is to prefix one's own behaviour against the whimsies of nature. Right now, there's only a peon in the office, and he himself. And yet, Cary is not able to accept the fact he could leave all his work waiting on this dry, rainless day. The damned rain! Muttering to himself, Cary walks across the room to make himself a coffee.

The peon comes in, 'There's someone here to meet you, sir!'

'Did you get the name?'

The peon holds out a piece of paper.

A frown takes shape on Cary's face. He knows that name. He wouldn't have allowed anyone else had they turned up without an appointment. But he can't turn this man away. What would he want? And from him?

He says, 'Show him in,' and adds another cup of water to the kettle.

'Good evening, Mr Cary,' says the man as he strides in. He doesn't wait to be asked to take his coat off. Clearly, he plans to stay awhile.

'Good evening, Dr Fitton. I've heard a lot about you. Of course, this is our first meeting.'

'I'm hardly a name you hear often, Mr Cary.'

'You're being modest, Dr Fitton. A doctor and yet you have devoted yourself to the study of geology. An esteemed member of the Geological Society, elected to the Royal Society too. But it wasn't good news when I heard of you the first time.'

'Oh, that! It should be overlooked, I was only eighteen at that time. I was studying in Dublin. I just took it upon myself to go looking at stones with a hammer, a compass and a bottle of hydrochloric acid in my bag. The British soldiers caught me. The Irish revolution had shaken up the whole of Great Britain at the time. Irish people were banned from carrying weapons of any sort. A hammer and acid in my possession and an Irish name to boot, what else could those rogues do but throw me into prison!'

'Ultimately, a letter sanctioned by the Trinity College authorities had to be obtained for your release. Am I correct, Doctor?'

'Absolutely. I was fortunate, Mr Cary. Not everyone is. In spite of friends and well-wishers, many accomplished people serve sentences in prison. Am I correct, Mr Cary?'

John Cary stands up and says, 'Some coffee, Dr Fitton?'

'Of course. How can there not be coffee when a British and an Irishman sit together over a conversation?'

Cary announces as he readies the cups, 'You have already made up your mind about this conversation, I see, without any notice to me! But had it rained, you wouldn't have found me here.'

'As I said, I am fortunate.'

'I'm aware. I too know a few facts about you. You've made a name for yourself in our scientific societies in a very short time. Your fortunes have smiled even more since your marriage. Your wife belongs to an aristocratic English family. Here, your coffee.'

Dr Fitton takes a sip, 'Delicious. Hazelnut?'

'Hmm…'

The room is suddenly full of silence. Fitton is totally attentive to sipping his coffee. Almost done with his, Cary straightens abruptly, gulps what is left and puts his cup down. 'What do you want to hear, Dr Fitton?'

Fitton's smile is a sliver. 'That which I do not know, which the entire British press has no clue about, which only you are aware of, that.'

'You are certain that I know?'

'I am, absolutely.'

'Oh, I see. Enlighten me then, Dr Fitton.'

'Where is William Smith?'

'You aren't a detective, neither are you a government investigator or police. Do you have the right to interrogate me?'

'I'm not here to interrogate you, Mr Cary. I'm here to let you know that we need that man. Not only for England or Ireland, but the whole world needs him. He is invaluable.'

'Is that so! And who may I ask has decided what he is worth, Dr Fitton? You, or the British Parliament, or the President of the Geological Society?'

'Think of it as me. Is it not enough that I, a respectable member of the British Geological Society and the Royal Society, Dr John Fitton, do not have an iota of doubt in considering Mr William Smith, the originator and original authority of the Earth sciences? Is that not affirmation enough for you?'

'Well, it does seem enough. It does seem like Smith's amazing fortune, doesn't it? A blue-blooded member of the Royal Society singing praises for a smith's son. Ahhh! If only he knew.'

'I do want him to know, Mr Cary. I've rushed here to find him. Where is he?'

'Why ask the same question over? How would I know where he is?'

'Only you would. His closest friend.'

'I, a close friend? To that luckless, mad vagabond? Would I have then allowed him to rot in prison for eighty days? Not visited him even once or try to pay off his debts?'

'Why haven't you, Mr Cary? That too is my question.'

'I ask myself all the time. And I make up my own answers. Metaphysical answers that may not please you at all. But do hear me out. Lord Christ had to bear the cross through roads full of onlookers even as he was being caned. Socrates had to sip poison in a dark cell. Galileo died, blind, coughing up blood. They had no close friends. John Smith's son, William, too, is without friends.'

'So, you refuse to tell me where he is?'

'I do not know. Because I am not his dearest friend.'

'Do you not wish to find out?'

'I no longer feel the desire to, Dr Fitton. And what good would it do? Just consider this scenario, we bring him back and he throws himself into the exploration of science once again. He discovers something novel and without meaning to, punctures blue-blooded pride again. We know how it'll end, right? He'll be humiliated and disrespected, what he deserves will be snatched away, and if it hurts their interests, he'll be thrown back into prison. It's better he stays untraced. If he is alive, let him live as he desires.'

'Should we slight a man such as him? The whole world acknowledges that the British know how to honour genius. How can we then neglect such a living legend?'

'What is it that you said, Dr Fitton? A living legend? How well we arrange our words, we, the educated upper class who flaunt our degrees. The world listens to us! We who always stand first among the civilized! I love how you speak!'

'You mock me, Mr Cary.'

'No, no, I do not mock you. Let me ask you a simple question since you are a respected citizen of this country.'

'Indeed. What would you wish to know?'

'Have you ever been to the British Museum, Dr Fitton?'

'Often enough. It is the world's best, abounding with scientific and historical artefacts.'

'Wonderful. The same British Museum acquired Smith's entire collection for a mere five hundred pounds. Thirty years of his toil, close to 2657 fossils from almost 693 species. Its price was determined by the British Treasury. Several wealthy worthies were willing to buy it, even up to a price of sixteen hundred pounds. But Smith refused to sell it to them. He was plagued by extreme penury and yet he returned often to the office of the British Treasury. After untold visits the government felt forced to acquire his collection, for a pittance of five hundred pounds only.'

'So I've been told, Mr Cary, it's unfortunate.'

'Ahh, please pay attention to this, Dr Fitton. What do you know of his misfortune! He was paid only a hundred pounds of his due, of which he used thirty to pay off his nephew's school debts. This is his sister Elizabeth's son, Phillips. When Elizabeth and her husband died, Phillips was only seven. A brilliant boy but the government has helped in no way for his education. Smith has taken the boy's entire responsibility upon himself. Of the rest of the money, he sent twenty-five pounds to his brother John. And the forty-five left over went to pay off his own debts.'

Dr Fitton is shocked, 'Dear God!'

'Now listen to what happens next. Arrest warrants are then issued against your brilliant mapmaker. His dear friend, Brogden, writes to request the British Treasury to clear the amount due to him. Brogden is a Member of Parliament, so the Treasury cannot ignore him. So, they send the mapmaker thirty pounds of the four hundred due. And that is how your legend just about avoids sleeping on damp prison floors in the chill of winter. He manages to avoid prison.'

'And then, Mr Cary?'

'You do seem curious. I'm enjoying it. Let me make you another cup of coffee. You will need it to continue with the unforgettable tale of your living legend.'

The unruffled Cary pours coffee into Fitton's cup.

'As I was saying, the biting cold passed at last. On February 12, Smith received a letter from the Royal Treasury. He would be paid another hundred pounds on March 28. To receive those hundred pounds, Smith walked mile upon mile to the Royal Treasury.'

'He was given a mere hundred pounds?'

'I'm not done, Dr Fitton. After he had stood in those cold wet corridors for two and a half hours, a cheque of seventy pounds was handed over to the mapmaker. This made a total of the previous thirty plus seventy now and the hundred pounds given at the time of taking over the fossils. Since then,

William Smith had not stepped into the Treasury offices ever again. Instead, he went to prison.'

'Unbelievable!'

'You work for the government, you should well be aware of its machinery, Dr Fitton. Although, I am convinced that had it been someone from the upper classes, the matter would not have gone so far.'

'I am truly sorry, Mr Cary.'

'Sorry about what? For that unfortunate soul, or for our entire tribe? Have you thought of what history will say about us in the future, Dr Fitton?'

'I don't understand!'

'What don't you? In a couple of hundred years, when people read of the history of England, an enlightened country will be revealed to them. A country that has illuminated the world with its science, a country that ignited an industrial revolution, whose sharp sword of logic has beheaded idealistic notions. Think of how inappropriate the chapter on William Smith will be in that book of history. Instruct your Parliament and Royal Society to destroy all documents that refer to William Smith. The world of the future must not know that there existed someone called William Smith, the Father of Geology. And if there was one such, he must have belonged to the upper classes, one of you that is. Why not try your best to erase the name Smith from our annals altogether.'

'You mock me again, Mr Cary. How does one erase a name that's spread across the land, whose work is seen across England till the South of Wales, from the canals of Somerset to the Estates of Hastings, Bath, the Irish Shelf, whose labour marks every street of London? Surely you understand that I have come here with a special request.'

John Cary lets out a silent laugh. He looks askance at the hefty man with a fallen face. This man hasn't come to ask after Smith, his business is something else. These people don't feel an iota of shame. He begins to put his bag in order.

'It's time for me to get home, Dr Fitton, please keep it short.'

'I…mean…' Dr Fitton falters.

'I do not have Smith's address as I told you earlier. I cannot help you to get in touch with him.'

'No, no, he doesn't need to be present. I've come to see the maps he drew.'

'Those maps can be found at all bookshops in Oxford. A printing press does not sell maps.'

'Not the ones that you printed a few months back. I'm referring to the yet unpublished maps.'

'What do you mean?' John Cary straightens up and stares at Fitton. He is livid. But Fitton has come begging today. He's not fool enough to cower at that face.

'You possess thirteen of Mr Smith's geological maps. I need them for special research. I simply wish to take a look at them, nothing more.'

'Sorry, Dr Fitton. William Smith has entrusted the maps to my care. My press alone can print and publish them. I cannot show them to anyone without Smith's permission.'

'Please, Mr Cary. It's my plea. I will come only for a day to take a look. I will sit where you give me room. I will carry nothing with me, no paper, no pencil. I'll study only with my eyes.'

'Were you looking for Smith only for this?'

Fitton continues to stand, his head bent.

'You people are incorrigible. But, good night, Dr Fitton. It's time for me to leave.'

'Mr Cary, I beg of you. Please do show those maps to me. I have no intention of copying them. But it's necessary to take a look at those maps for the sake of science.'

Cary suddenly starts to laugh. He's abandoned all sense of propriety and is guffawing. As if a stupendous comic act is taking place in front of his very eyes. Fitton's face has turned red.

'Forgive me, Dr Fitton, I'm being very rude. I should collect myself. Look, my laughter escapes me still. I am sorry, so sorry.'

John Cary continues to laugh. There's nothing in his behaviour to indicate that he is indeed sorry. Fitton continues to stand, his sighs held at bay.

'Since you have come for the sake of science, I promise to keep your request. But you too must keep a promise for me, Dr Fitton. This too is for science's sake.'

'Of course, Mr Cary. Do you mean about finding Mr Smith?'

'I've known two scientists from up close, Dr Fitton. One, you, and the other, William Smith. You both explore the same branch, yet you both are so unlike. That man gave only half-answers to the thousands of questions posed to him. He became almost an alien even as he continued to stare at stones. He observed and took notes, that's all. His reasons for each observation took him months to detail and define it. He spent years together doing this. A foolish scientist, that's him. Today, I've met a shrewd scientist. One who answers before he's questioned.'

'I'm sorry yet again, Mr Cary. But please complete what you must say.'

'I've said it before, and say it yet again, I do not wish to look for William Smith. But I'll ask you my initial question again. Have you been to the British Museum, Dr Fitton?'

'I've answered that already. Not once, but I've visited it on several occasions.'

'In that case, visit it one more time. Try and fathom the history of what the museum or your Geological Society has done with William Smith's collection. I need a detailed explanation. Find out about each of their moves and come to me. On that day, I'll show you those thirteen maps. Then you can stay as long as you wish. But do come with these details. This, too, is for the sake of science. Good night, Dr Fitton.'

YEAR 1794, SOMERSET

It is Christmas and the miners have been given two weeks off from work. Barring five or six of them, the rest have been given permission to visit home. Of the unfortunates who are staying back, one is Nash. If one looks carefully at this youth, perhaps twenty-three or twenty-four years of age, one can make out that there's Asian blood in his veins, although he calls himself British. Nash's parents are no longer alive. He's grown up in an orphanage. He's ill-tempered. And almost no one gets along with him.

Perched atop a large rock, Nash is chewing bread. Dry bread with a bottle of water. Spotting him from afar, a young girl tries to clamber up the rock. Her name is Oliva. The daughter of the High Littleton Coal Company Sub-surveyor, Brookes. Although she is past eighteen, for some reason, Oliva is still unmarried. Not bad to look at, her face is oval, her hair tawny, and her eyes, tiny compared to the rest of her face, and squinty. Usually, this would not be a problem in getting a girl married. But rumour has it that Oliva is a little loose in the head. Some say Oliva's mother is stark raving mad. Nobody wants to get married into a mad house. One hardly sees Oliva's mother outside. Sometimes, at night, loud shrieks can be heard coming from the Brookes's. Din of fights and skirmishes in the worker colonies are a daily affair. No one bothers about such things. And then again it is difficult to say whether it is the mother shouting or the daughter. In any case, Brookes's house lies almost at the edge of the workers colony.

Nash has noticed that in the last few months Oliva has tried often to talk to him on her own, asking after his health, even offering advice, unasked for. Nash does not know if this is a good sign or not. He likes to think of his own past often, and company irritates him then. Those who he is most mad at are his birth parents and there's no way now for him to get a hold of them. Nash is mostly distracted in these moments and if someone tries to break into his thoughts, he becomes incensed.

There's a strange charm about Nash, a mix of the Asian with Anglo Saxon. His eyes are elongated, their pupils dark. A full moustache. His chest broad and muscled. The only drawback, Nash does not smile. No one can say what he would look like if he did laugh. His face has frozen into a hard, dark look.

When Oliva finally manages to climb up to Nash, she is panting. Nash gives her one look and goes back to chewing his bread.

'What? Only bread? No soup?'

Nash does not respond.

'It's festive season, why only bread? Here, take this.' Oliva extends a clay pot. Nash takes a look and turns back to his bread again.

'Take it, I say. It's trotters soup. Try it.'

Nash answers, 'Give it to the one you brought it for.'

'Who? Father? He's had it already. This is my share. I've brought a large portion.'

'Then sit and drink.'

'Of course, I will. I told you, I've brought extra. Here.'

Finally, Nash tosses the last piece of bread into the bushes nearby. 'I'm done already.'

'What are you? A man or something else?'

'What do you mean?'

'Well, you do look like a man, but you behave like a pig, its teeth bare. Grunting all the while.'

'Get on with your work now, will you? I can't stand your blah-blahing.'

'You don't like a thing. Why even live with so much anger stored up? Drown yourself in the pond instead.'

'It's my business what I do.'

'Not yours alone. If you behave like an animal, you can't live with human beings.'

'I'm leaving.' Nash makes as if to stand up.

'Don't you dare. Sit down now. You're full of rage. Shut up and sit down. You may leave after I've had my say.'

Nash plops down again, a tad taken aback. Had this been a man, he would have pushed him aside. But she's a woman, her arms akimbo, parked across his path. He cannot lift his hand on her.

Looking away he says, 'Tell me quickly then.'

Oliva plonks herself down, next to him, smiling slyly. She picks carefully at the grass around, slim long ones and puts them down on a stone next to her. Narrow stalks of grass, all the same length. Once in a while, she chews on one. Forget finishing up instantly, she seems far from the beginning of the conversation.

Nash was agitated at the start. But now he is calming down. His temper's like an iron rod, quick to heat, quick to cool.

A lock of hair quivers gently at Oliva's left ear, like pendulous flowers on a plant. Her neck alabaster, and a bunch of golden-brown hair rests against it. Nash's voice dips as he stares mesmerised.

'You wanted to say something.'

Oliva turns her neck and laughs with amazing grace. Nash looks away, swallows hard. He's heard that women's laughter can ensnare. He's seen many such laughing lassies at the pubs.

But they were loud, shrill almost. This is not that. This makes his insides shiver, like never before. He doesn't know if this portends good or bad.

'Won't you go home for Christmas?' Oliva questions.

'Nah.'

'Why? Has your wife bolted?' She's not mocking him. She's earnest.

'I have no one.'

'Then what's the point of such anger? Everyone goes home in the holidays. You have no home, hence no holiday. Although, we too aren't going back home this season.'

Nash finds this curious. Sub-surveyor Brooke has in fact been nicknamed Holiday Brooke by all of them. Because, almost every month he is on leave for at least a few days. He's also very reluctant to get back when he leaves for Easter or Christmas. The same man is now not taking leave, very strange indeed.

Oliva turns to Nash, 'No surprise there. My father's given up his leave willingly.'

Stranger still! Holiday Brooke has willingly forgone his leave. Unbelievable! Nash blurts out before he can stop himself, 'Why?'

Oliva giggles, 'It seems your new surveyor is slightly loose in the head.'

Nash's expression hardens, 'Has your father said so?'

'Why just my father, everyone else, too. It seems the man doesn't survey much. He keeps going up and down the shaft, staring at rocks.'

'Well, he has come to check out rocks.'

'Why would he come to look at rocks? He has come to figure out the depth of the coal layers. Stop teaching me survey work,' Oliva says this almost as if nobody knows surveying better than her.

Nash laughs.

'Dear me, you can laugh. Phew! What a relief. The man's been named Strata Smith. Now, what could strata mean?'

'Layers. There are different layers of rocks under the soil. Coal is one of these.'

'A young man, with no friends, no relatives, no addiction, no desire to have fun, lives in a haunted house and stares at rocks. Quite like you, though, right?'

'I don't stare at rocks. Are they things to stare at?' Nash retorts.

'That's even worse. At least there's a method to the man's madness. He's happy staring at stones and rocks. But God knows what pleases you. And yet, it seems you are both similar.'

Nash suddenly recalls something from two weeks back. Surveyor Smith was coming down an iron ladder. He had stopped all at once. Switching on the light of his helmet he had begun peering around. Below him, coal was being dug out. Sub-surveyor Brookes grumbled, 'Darned problem this. God knows what he keeps staring at all day long. He's come to look for coal and spends the day looking at rocks instead. Nash, go and summon him. Ask him to hurry up and come down immediately.'

Nash had climbed up unwillingly, 'Come down, please, sir. Mr Brookes is waiting for you.'

The surveyor had looked down only after he'd been called several times. He signalled to Nash to climb up. Nash had stepped up the ladder, but the more he went up, the more the surveyor climbed up too. Finally, at the mouth of the mine, almost hanging down loosely, he had called out again, 'Come Nash, stand by me.'

Nash had been taken aback. How did Surveyor Smith know his name from amongst so many workers? He'd climbed up and stood, somehow, one foot balanced on a two and a half feet iron rod. That rickety thing had almost been like a rope stepladder. But iron, instead of canvas.

The surveyor took no notice of anything. He said, 'Look at this carefully, Nash. There's loose soil everywhere. And a thick layer about eight to ten inches of pebbles and gravel mixed with moss. Not tightly packed, but a layer that crumbles rather easily. After this, a layer of red stones. This won't crumble, it's hard. This is marl, a type of limestone. If you move eastward from this, you'll find this layer inclining down that way. Under this is a layer of greenish mudstone. This too inclines eastward. This means two different layers, that have hardened one on top of the other. Now come down with me, a thousand feet or so. Look around carefully where I stop. Come.'

Surveyor Smith was climbing down. Nash followed. Smith paused and asked, 'Look carefully, what is it that you see?'

Nash dithered, 'The colours have changed?'

'Very good. Not only the colour, the rock type too. This is sandstone. Check its grains, its sandy.'

Nash had rubbed his fingers across the surface, 'Yes, sir, it is.'

'Anything else?'

'The layer's thick, deep too.'

'Excellent! What else?'

'There are layers of coal between it.'

'You're making good progress. Tell me more.'

'There are cracks across it, almost as if someone has used a hammer.'

'Fine observation. Although the hammer wasn't used by men.'

'Who then, sir? It can't be dogs and cats!'

'Ahh, don't be foolish.'

'That's what, sir. Then who?'

'We'll discuss that later. Tell me first what you see on both sides of the cracks. There are strips of varied colours on this rock. Keep your eye on one such strip and move slowly across to the other side of the crack. Anything you find interesting?'

'Not much, although it seems that across the crack the strip has shifted downward.'

'Wonderful. Not shifted, it's called faulting. Anything else?'

Nash had sighed, 'Nothing else, sir.'

'You've said enough. Now listen to what I say. Look, the top here is not as smooth as the sandstone or limestone. This is like someone has compressed two sides of a carpet. It's folded. Wavy folds and cracks in between, isn't it?'

Nash couldn't see much. He was panting from the strong headlight and the hot air blasting from the cracks. He somehow managed a, 'Must be, sir.'

'Not must be, it is. Something has pressed on the two sides of this layer, Nash. It's folded up from the force. And cracked too. The layer on this side has ridden up to the crack. Although, look, there's no sign of force on the red and green stones.'

Nash offered, hesitating, 'I've worked here four years, sir. The shaft in this mine was started about two and a half years back. I haven't seen anyone exerting any force in this time. There's something you can do, sir. Uncle Peter has been working this area for the last twenty years. We could ask him about this.'

Surveyor Smith had chuckled. 'Do you think this is a man's doing? Who can have the power to burrow thirty feet deep and exert such force? Try to fold a piece of rock and see if you can. This is the doing of an infinite power, some boundless force.'

Nash's face had taken on a glow. He knew what the surveyor was hinting at. He had added, 'I know what you mean, sir, this is God's work! In rage he must have touched these rocks and they folded up.'

The surveyor turned his enraged eyes upon Nash. Then smiled faintly, 'When there are no answers, one turns to the name of God. Who has done this and why, what was that force, did this happen only in England or across the world, no one wants to ask. No one wants answers. So, it is best to put it all on God.'

Nash was at a loss. He felt terrible looking at the surveyor's expression. There was pain written all over it. The pain of an unanswered search. Just like Nash's pain when he searched in his heart for his parents. If only he could find them once, he could set some things right. But where should he begin his search, Nash has no clue.

Nash had remained silent. Like him, the surveyor was also suffering helplessly.

'Why do they not wish to see that this is not possible in six days. Not even in four or five thousand years. This is far more ancient. This has been destroyed and created over and yet over again. Massive twists and lurches have caused a deluge to sweep over all this. A tremendous force has washed over and the layers have gone out of control, the top sinking to the bottom, the bottom floating up to the top. And then again calm prevails, everything is born anew. This is the cycle of creation and destruction. These rocks can never only be four thousand years old.'

'How old do you think they are, sir?' Nash asked and immediately bit his tongue. He sounded too much like a scholar. The surveyor might make fun of him.

But the surveyor had not. He'd turned to Nash with a strange indefinable look. What was it about those eyes that had paralyzed Nash. The surveyor's eyes seemed to burn in the light of the helmet. How far had his vision penetrated? He seemed to see a universe created before, much before, than the one created by God. Or so Nash felt. A chaos seems to have been unleashed.

Nash has been desperate to know who his parents were. But the separation had happened just twenty-five years ago—at his birth. What was the surveyor seeking? Perhaps, he will never be able to give up this search, even if God were himself to stand before him. What is this Nash is thinking? How can there be anything older than God? His head is all aflutter now thanks to the surveyor. If he keeps his company much longer, he may even forget his search for his parents. This is a man who is looking for the one who has birthed the universe, why should Nash take him on? God alone knows what madness the surveyor has got himself into.

The surveyor spoke in a whisper, 'No one asks this question ever, Nash. How old is the world? Who is its creator, its destructor, the one who creates it again? No one wants to find out. But we must, absolutely. Oh! The pain, the pain. How long will I live, Nash? At best thirty or forty years. Will I be able to get to the bottom of these questions?' The surveyor could not stop his tears.

Nash had felt tearful, too. Like a father consoling his weeping child, he said, 'Don't be anxious, sir, if not you, someone will, soon enough.'

'When? Will I live to see the day? I so wish to know, Nash, when did all this take place? Who is that power? This is not something that happened five thousand years ago. And, therefore, I do not hold it to be God's work.'

'Shh, sir, be careful what you say. If anyone overhears, they'll dismiss you, wreak havoc in your home. What must one do with you! Come down, sir, please. A layer of coal has been found, it needs to be measured.'

Would he be able to share this incident with the girl? She'll probably not make any sense of it, instead parrot other stories to him. The poor surveyor's reputation will be in the balance. Nash changes the subject, 'Why did your father himself give up his leave?'

'Don't even ask. The Company is facing severe issues with the new surveyor. It seems he does no work. He simply wanders around from this mine to that, stares at rocks and stones, and keeps taking notes.'

'Not at all. No one before him has ever measured the layers of coal with such detail. Where it begins, where it grows thick, its varied types, each point goes into his books. The previous surveyor did nothing of the sort.'

'Ooh! You know so much. I've heard the man is illiterate, has taken no exam at all.'

'It's not the same, taking an exam and having knowledge. Look at you, spouting such theories, and you said you know all about surveying. How many years have you spent in a school?'

'Hmm, I see. The man has bewitched you. I could understand had he been a woman, but to be under the spell of a half-mad man…tut-tut…what is one to say!'

'Say nothing, since there's no one to worry about me. I'll be on my way now, the bell's about to go off.'

'No bell today, mister, it's a half day. Tomorrow's Christmas eve. A market's been set up in the next village for the next two days. A lot of stuff is on sale. Most of the workers are headed there, their pockets warm with bonus money. Haven't you received your bonus yet?'

'I haven't taken it yet. Let me go look for the surveyor.'

'He's not the man. It's my father who doles it out. Is the Company silly enough to hand over such a large amount of money to that insane fellow?'

'I see.' Nash stood up.

'Why not come home tomorrow, please? Take your due from my father? And go to the market after. I could join you if you wish. I need a few knick-knacks, a new hat too.'

'We'll see. Although I don't really need anything. And that which I do, I find in the pub every evening.'

Nash clambers down.

'You!' Oliva calls out. Nash stops, but doesn't turn around.

'Can you introduce me to the surveyor someday, your Strata Smith?'

Nash waves his hand, 'No point in that. No human can do anything for him whose mind's been invaded by stones.'

Oliva is furious at this but stops herself from saying anything. She likes Nash. He's not like the others. She's curious about these men, men who are different. She's keen to know what makes them tick. She's consumed by her curiosity these days. This too is a search.

Instead of heading to the worker's colony, Nash goes toward the surveyor's house. Oliva is not wrong. It's a huge house, standing silent like a ghost. The front entrance is extremely well-kept. The hedges have been trimmed. A high portico encloses the building. The surveyor is sitting out, scribbling away.

Nash stands by, shyly.

Quite some time passes before the surveyor notices him. 'Oh, you! Haven't you gone home?'

'I haven't been given leave.'

'How's that? Hardly any work will happen in these two weeks. Go, go visit somewhere.'

'I have no home, sir, no father or mother, no sisters or brothers. Where I am is home.'

'I see. Take a look at this drawing then, Nash.'

He holds out a hand-drawn sheet for Nash. A horse-drawn coal cart inside a mine, miners piling it up. And two men pulling the horses forward.

'This's a fine piece, sir. A boy named Francis who works in the nine-inch coal seam had also made such a drawing, sir.'

'Francis? Really? That means anyone can draw like this, right?'

'Of course, sir. Anyone with a fine hand can, sir. You have a good hand, too, sir,' Nash stops even as he says this.

Strata Smith's face is suddenly downcast. Has Nash said something hurtful?

'What I meant…was…you are a fine artist, too, sir. You could draw even better if you wanted.'

'Draw? No, no, I do not want to draw pictures. I have no wish to be an artist. But I want to create something, something very different.'

'Why not then draw stones, sir, the rocks in the ground beneath? Just as you showed them to me the other day. The ones that incline, those that fold up, all of these. If you need colours, I'll bring them. If only we mix those rocks with glue, what amazing colours we will get.'

Smith looks at Nash, his eyes unwavering. The same dream-drenched look from earlier, when Nash had felt afraid while on the stepladder inside the mine. But he is no longer afraid anymore. He's fallen in love with his mad master. When two lost souls, completely alone, bond together, it's a play of unseen fate. It births intense attraction. Nash desperately wishes to somehow wipe the sweat off his master's flushed face.

His voice is steeped in affection when he says, 'You haven't had a bite, sir. Should I buy bread and meat for you?'

'I was hungry, but not anymore. You don't recognise what you just suggested, Nash. To be able to draw onto paper what lies beneath the earth! To delineate this tri-dimensional world in two dimensions. One draws easily what lies above the soil, the rocks, the rivers. But to measure each layer, to map where they start and end under the surface, I cannot even begin to imagine this wonder. Nash, is it even possible to draw a map such as this, will I be able to?'

Why do his eyes tear up again? The cantankerous and rough Nash softens now just as mudstone would when it soaks up the rain. He has never been washed away like now under the force of emotions. Who has entrusted him with the responsibility to love this graceless madman? This is how our duty to the world is handed over from one to another. But this man is his master. He must be prudent in his love for him. His is one-sided love. The silly man must never know.

'You will be, sir. How is it that they will not return your love, these who you live for, for who your night and day become one? These stones will speak, sir. No one understands them, only you. Pass on to us what they have shared with you. Who can speak of the future? The world outside our High Littleton village is populous. There will be a wise man among them who will understand you. Write what you consider right. Someone or the other will soon look upon the world with your eyes.'

The man is overjoyed. The eyes which were helpless and despondent a moment ago, now light up. Nash smiles broadly. This is perhaps a first for him too, he feels happy inside, light and floaty.

'May I say something, sir?'

'Hmm, yes,' Smith looks back again at the sketch.

'The Christmas festival begins tomorrow. Can we not begin the drawings from the day after? Let's go celebrate today, just up to the pub. Would it hurt to have a night's fun? We'll dedicate ourselves to the earth's insides after that. What say?'

Nash says this but repents instantly. The surveyor is his boss. Would he even consider going out with a hired hand such as him?

The surveyor suddenly walks away, goes inside the house. Should Nash leave? This is a gentleman, unable to refuse him to his face. Perhaps it would be correct to leave immediately. He's crossed his limits. Nash shuffles toward the gate.

'Nash, where are you off to?'

Nash turns around, startled. The surveyor has emerged again, clad in a thin shirt, a turquoise greenish-blue shirt coloured like the waves of the ocean.

Nash can't help but smile, 'You must get a coat, it'll be a chilly night.'

'Not tonight. Our bodies will keep warm with what we pour inside.'

Nash guffaws.

Nash hums as they walk, a song a boy in his orphanage often sang, "The Beggars Chorus". A song to celebrate the lives and trade of beggars.

And a begging we will go will go will go
And a begging I will go.

...

A bag for my wheat
And another for my salt...

Strata Smith is jiving along. Nash snaps off a dry branch and plays to a beat on a rock...tra, tra, tra.

Smith now thumps his foot to the beat. This man, his master, is totally oblivious to how another person has come so close to him and been transformed completely. Nash looks up, his eyes dripping affection. God alone knows where this madness will end.

When two people of similar ages but widely disparate in their ways of thinking and learning get into a serious discussion with each other, it manifests itself in varied stages of hilarity. Almost at the outskirt of Rugbourne village in a lane in front of a three-storeyed house, two such people can be found chatting away every evening till it gets very dark. Today, Sunday, is a holiday. The two men have just returned from church. Neither had been keen to attend mass till the end, although one more in a rush than the other. Once back home, the man unrolls a huge map and places it on the floor, all the while straining to make the other grasp its details.

'Look at this carefully, Nash, this is the layer above the soil, made of grass and pebbles. Below this, red stone, and below that, green stones. Still below, a layer of golden stone, then ashen mudstone. And then sandstone punctuated with thick and thin layers of coal.'

'Where did the golden rocks come from, sir? We didn't see them the other day.'

'I couldn't show you then. It was embedded a distance away.'

'Is it gold?'

'No, although it does look like gold from afar. In the country where Mr Hastings lived, they build forts with this stone. Golden forts.'

'Who is Mr Hastings?'

'I can't explain all that now. But the country is called Hindustan. Across the oceans, far away from our England.'

It sounded sweet, our England, and yet Nash is compelled to ask, 'What is England, sir?'

'The country you were born in, where you live.'

Nash doesn't know where he was born. He likes to evade all references to it. Immediately he pipes up, 'Of course, yes, I know.'

'But we have a problem, Nash.'

'What's that, sir?' Nash questions anxiously.

'These rocks underground, are these confined to the soil of High Littleton or do they spread across from England to the mountains of France?'

Nash cannot fathom, which is England, what is France. He answers gravely, 'From what I see from your drawings, everything ends two inches under the soil.'

'No, no, you goof. Look at this scale. One inch here equals two hundred feet underground. Sketched longitudinally. This last layer, it's really six thousand feet under the ground. If you dig straight through the mine, you will see these. Does it make sense?'

There is indeed much to learn, but Nash does not wish to rile up the scholar in front of him. And then, since he's sharing his knowledge with Nash he clearly considers him worthy enough. Nash doesn't want to ask unnecessary questions and prove him wrong. So, he says very seriously, 'Hmm, of course.'

The man looks happy, 'Wonderful. Now tell me, why are the layers of coal here named as they are? Sleeving, Dangri, Kingswood. And then Number Seven, Number Eleven, Nine Inch. How did these strange names come about?'

Nash knows a little about this. He's heard of it from the old mine workers. His face glows now. 'Some surveyors and then some workers, too, were responsible. Mostly the ones who were dull. The witty gifted ones would name them differently.'

'Like how?'

Nash delights in Smith's enthusiasm. 'For example, Number Seven and Number Eleven. You'll know instantly that the men must have been as dry as dust. Used a number and finished the job.'

'Ha, what's this you say?'

'Now, Dangri. I've heard the surveyor had a mistress named Dangri. Her father worked in the mine here. The man loved his woman very much. And named the coal after her. And then Kingswood. There was a brewery at this location. It was closed years ago. One surveyor was a regular there. And, therefore, Kingswood.'

'Ahh. Quite something that.'

'Why don't you too name a layer, sir? You did find a layer the other day, do name it. Name it after the one you love. You could bring her back here after your wedding, she'll be delighted.'

Now Smith banters with Nash, 'Yes, of course, she'll float with happiness...won't she...off now, you oaf! Go, get me some coffee. The fool wishes to see my bride. Go, run now.'

Nash dashes in to get the coffee.

'Listen.'

Nash turns around.

Smith says, 'You mentioned a person I love, right? Well then, I name my layer Nash.'

The surveyor has turned back to his drawing. Nash is staring at him, rendered powerless. What did his mad master just say? So he does understand how much Nash cares for him. Or did he only utter whatever came to his head? Nash wishes he could tousle the man's sun-dried gold hair. Does he feel what Nash feels? Nash cannot believe so.

The sun's light casts Nash's shadow upon the ground. Smith surveyor notices it and calls out, 'The rascal is still standing here. You scamp, scoot now.'

YEAR 1795, SOMERSET

There are only two guests for breakfast this morning at Lady Jones's, the leading figure of the High Littleton Coal Company. This is unusual. Most parties at the Jones Estate are held in the evening. Those present include illustrious personalities beginning with Members of Parliament to owners of large mines, barons, earls and knights. A breakfast meet indicates a highly confidential discussion. This morning's guests are well-known, wealthy members of the Parliament as well. The lady of the house has business deals with them, and those deals have to do with the Somerset coal region.

Lady Jones calls the men by name, although they are high-ranking. This is because Baronet Jones had been responsible for mentoring them each step of the way into the business of coal. One is Henry, the other, Paul.

As they serve themselves, Paul begins, 'I just can't fathom what Somerset's plight will soon be.'

Lady Jones smiles archly, 'Somerset has never ever been in an enviable condition, Paul. I've seen it since I was married here—a miserable, coal-coated land full of pits and holes. Have you ever seen it better than that, Henry?'

Henry answers instantly, 'No, of course not, he didn't mean the natural collapse of Somerset, he was speaking of us, the mine owners.'

Lady Jones responds, 'I see. I wondered for just a moment whose company had suddenly changed a seasoned businessman like Paul into a nature lover.'

Paul hesitates as he says, 'No, I'm not referring to its natural condition, not at all.'

Lady Jones motions for him to stop, 'Look, now, come to the point instead of circling around. I understand only coal, and its layers. Because that is what my Lord Baron trained me in. Tell me immediately where the real trouble is.'

Paul answers, 'Duke Bridgewater is building a canal. I'm sure we realize what the outcome will be as soon as it is ready, right?'

Lady Jones counters, 'What could possibly happen? It'll connect Avon with Kennet and move on to the Thames. His coal from Wales will flood the markets of Bristol and London.'

Henry looks up at Lady Jones with admiration. Baronet Jones had treated them both like his sons and taught them the intricacies of coal with great affection, holding nothing back, giving them his all. But what he could impart to this woman, even as she stayed confined within the interiors of her domestic universe, is immeasurable, its sweep boundless. Or perhaps, it is the disciple's panache, far greater than her tutor's, that accounts for this. For a woman who has never stepped out of the limits of her company estate, where does her foresight emanate from?

Lady Jones continues to speak, 'This had to be, Paul. After continually supplying the country with coal for the last thirty-five years, Somerset today is all but drained out. Its coal is outstanding, but it has to be drawn from hundred thousand feet underground. There are cracks everywhere and an entire side along these gaps has sunk in as well. It's clear that this has happened in most of the best layers of coal. It would easily be the most profitable if we could sell this in the market. The Duke's mines are comparatively younger although they are more productive. But his problem is he can't tap the London markets easily only because of lack of transportation channels. However, once this problem is taken care of, it's profits all the way for them.'

'Shouldn't we do something then?' Paul asks.

'Find a new spot. Somerset cannot continue for long now. Its stocks are almost empty. You must look elsewhere.'

'That will not be possible. You are a rich aristocrat. You are far more affluent than we are. We have sunk in all our wealth into Somerset, we will most certainly starve and perish, ma'am.'

'Confound it, Paul,' Lady John lashes out exasperated. 'I despise being addressed as "ma'am". I am Lady Jones, the Baronet Jones's wife. I consider any other title an insult of high order.'

'I'm sorry, Lady Jones.'

Henry adds his bit now, 'Dear Lady Jones, we've come running to you in desperation. You must help us. Is there no way to protect our mines? Are we to lose our savings as well as our lives?'

Paul butts in, again, 'Had Baron Jones been alive would he sit tight having witnessed our plight?'

By the time Henry has attempted to nudge Paul into silence, the words are already said. Paul's irascible. When incensed, he can never hold himself back.

But Lady Jones chuckles. 'Had my husband been alive, he'd have made the two of you roast in the sun for hours and then chastened you! Fools, he would have said, who made pea-brained idiots such as you mine owners? Can you not even figure out that you must dig a canal if you wish to save Somerset? Sit together. Implore the Parliament that a canal is a must to keep a profitable mine like Somerset up and running.'

Henry is astounded. 'What are you suggesting Lady Jones! Should we build a canal? Our very own? But will the Parliament concede our request?'

'Why ever not? The need for several canals has been considered under the Canal Act. If Duke Bridgewater can get his own Coal Bill passed, then why not you? And if you can't, then sell whatever you possess and return to farming. Give up your seats at the Parliament as well. You are an incapable lot.'

'But how would this canal be made? Where will it begin and where will it head?' Paul questions again.

'Go bury your useless eyes in the mine pits right away. You seem to have forgotten all of Baron Jones's training! Have you ever looked at the maps? Or are you content having left everything to the surveyors?'

'No, no, we will take a look at the maps immediately,' Henry stutters embarrassed.

'You must, but let me tell you this. I learnt from my husband to study the maps for at least a couple of hours every day. He must have taught you this as well.'

Paul and Henry, suitably chastised, sit with their heads down.

Lady Jones starts again. 'The canal should be cut around Camerton. And it should finish at Limpley Stoke, right where the Kennet and Avon canals cut across. This new canal must simply join these old ones. After that, it would be enough to let the Kennet-Avon canals take care of transporting the coal.'

Paul and Henry pipe up as one, 'Unbelievable!'

Lady Jones looks at the floor now, 'And, sorry, if you want to ask about how the canal should wind its way, I have no answer. Because, I am not a surveyor.'

Henry counters, 'Respected lady, today I am proud to acknowledge you as the best student of our master. You had clearly already thought through all these matters. We're immensely grateful for this. No words are enough to thank you.'

'Alright now, there's time enough to thank me. Let's discuss the real issue. We have decided to build a canal. Right? Consider that the Parliament accedes to our request. In which case, who will build it? The soil here is not at all packed or hard. It's a job, then, only for an exceptionally skilled engineer.'

Paul's suddenly all excited. 'I know just the man. Engineer Rennie.'

Henry queries, 'Isn't Rennie the same who's constructing lighthouses, bridges, tunnels, one after the other?'

'The same. He's also a member of Parliament,' Paul's eyes are all lit up.

'Yes, dear boys, he's truly fitted for the job. All he needs is a pair of scissors, to run across the land.'

'Meaning?' A flabbergasted Paul's not sure he's understood a word of this.

'Meaning that Rennie is so supremely busy that he won't be able to spare even five months for this job.'

'Have you already discussed the matter with him?' Henry forgets to blink, so dismayed is he.

'What else? Should I wait for light to dawn on you two fools?'

But Lady Jones's blazing eyes soften suddenly. 'Somerset was always my Baron's dream. I was married at age twenty. Do you know where he brought me for a honeymoon? To this arid Somerset. People recall their honeymoon as a bed of roses, and my destiny was this bed of coal. He would take my hand and walk me from one end of Somerset to another, teaching me about stones, helping me learn to differentiate black coal from black stones. My friends told tales of honeymoon swims in blue seas, and I swam through dark mine holes, showered in coal dust. I couldn't sleep nights, crying through them hunched on portico chairs. When tears streamed down my cheeks, they ran like grimy rivulets. I'm sure your wives do not live here with you, do they?'

'No, in London. We travel up and back.'

'He never ever wished to leave Somerset, although Bristol was just ten miles away. One day I picked up a fight and drove off. To Bristol. My father had bought me a house there.'

'And then?'

Lady Jones turns to look at Paul. She smiles, 'What then? Wisdom dawned. My man was a diamond, embedded in coal. I'd sin if I hurt him.'

'If only our wives felt the same.'

'They will, in exactly the same way if you begin to pursue your dreams just as Baronet Jones did. Enough. Let's get down to business now. Engineer Rennie says that were he to be given an able assistant, he could take on this job.'

Paul grunts, 'Dear me, this is as complex as Archimedes believing that he could move the earth. All he awaited was an ideal spoon, and he could have done it.'

Lady Jones rings a bell twice. A man walks in. He greets her, bending very low. It seems as if an oval egg-shaped bald ball is attached to his frame.

'Mr Ashley, you do recognise these gentlemen, don't you? They've frequented this house as young men. Of course, now they are fathers themselves. Members of Parliament, Mr Paul and Mr Henry.'

'Good morning, sirs,' Ashley bobs up and down, twice, like a wound-up toy.

'Please fetch the envelope lying on my table. Bring it as it is. Please do not read the sender's name.'

When Ashley hands her the envelope, Lady Jones takes a letter out of it and passes it on to Henry. It's addressed to her.

'Respected Lady Jones,

I've been taking care of the High Littleton coal mines for several years now. Its productivity is no longer as it was in the past. We need to have scientific and well-examined facts if we wish to dream of profits from it in the future. I could undertake this job should you wish so. There will be no payment expected at this end. Although, I do have a condition. You would have to accept sharing a part-ownership of the High Littleton coal mine with me. If you agree, I could start work at the earliest.'

'Who is this?' Paul raises his brows.

'The name and designation are mentioned at the bottom of the letter.' Lady Jones is exceptionally calm.

'Is the man well-endowed?' Henry looks up having read the letter.

'Not at all. He lives in a dilapidated house that costs him very little. He sups at the dirty colliery kitchen.'

'Well, for sure, he's daring,' Henry comments.

'True, undoubtedly audacious.'

'He's neither brave not tough, just mad. Why else would he wish to become a shareholder in a survey company?' Paul butts in.

Lady Jones smiles gently, 'You both are right. The man is as mad as he is brave. And you require just one such as him. You both will get the Parliament to pass the Somerset Canal Act. But prior to that you must get Engineer Rennie to agree to the job.'

'What if Rennie does not agree? Would he, without speaking with this man?'

Lady Jones stands up. She's ageing. She uses a stick to walk. It is difficult to fathom that she is past fifty when one looks at her from behind.

'I have nothing more to say now. From here on, it's up to the two of you. We've been discussing this for more than thirty minutes and I need to get back to my office again. Goodbye.'

YEAR 1794, SOMERSET

It's a Sunday once more. Time again for some afternoon banter, between two friends of the same age but unequal wisdom. Today's topic of discussion is fossils.

'Do you know, Nash, that each of the many varied coal layers here such as Thornberry, Globe, Kingswood, Number 11 and others, is home to different types of creature fossils? In some we see remains of sea creatures, in others, the shapes of river beings. Surely this means that in the same area at any one time there were oceans, and then it turned into land.'

'What magic is this, sir, where land turns to water and back to land again?'

'Sleight of hand indeed, but why, that I cannot figure out.'

'The priest had once told us about Moses, the messenger of God, the one who walked on land by parting the seas. This may have happened then.'

'What proof is there that Moses walked the seas?'

'The priest said so. What greater proof than that?'

'Nash, let this talk of church be. Remember this always, do not believe what you haven't seen with your own eyes. Understood?'

Nash nods vigorously, harder than he needs to.

Smith takes his eyes off his maps to stare at Nash, 'What are you doing?'

'I won't believe it, no, not at all.'

'What won't you?'

'Oliva! She doesn't eat her own share, but saves it for me.'

Smith looks surprised, 'Oliva. Isn't that a girl's name?'

'Yes, she's a girl of course. That is, she does look like a girl, but behaves just like boys. Keeps bullying me all the while. Quite quarrelsome. Sometimes she screams so loud I can barely breathe.'

'A quarrelsome girl who brings food for you! Hmm, it's beyond me.'

'Even I think so. That's why I said I won't believe what I haven't seen.'

'Let that be now,' Smith turns back to his sketch. But a little later he thinks of something and turns back to Nash again, 'It could be true though.'

Nash's taken aback, 'What, sir?'

'Perhaps the girl brought it for you without eating any. Maybe she's fallen in love with you.'

'Nonsense, how can you even suggest it? Love! With me?'

'That's true. Who can love a rascal such as you? You and I, the only two useless souls of this world! Which girl can love us?'

'She asks about you as well, sometimes.'

'Really? About me?'

'Yeah, asks about this and that. Says often about wanting to meet you.'

'Goodness now. Does she intend to choose one of us? I'll be sure to lose then. My skin's baked by the sun, my hair ashen, and no social graces at all. No, no, do not bring her here. And you too, don't blabber thoughtlessly.'

Nash nods vigorously again, 'I won't.'

Smith goes back to his drawing. And then again, looks back at Nash, 'It's quite possible, I say.'

Nash looks at him timidly, 'What, sir?'

'It could be love. I mean, maybe she loves you.'

'Nonsense, how's that possible?' Nash mumbles shyly, mortified. Smith's enjoying it.

'Is Mr William Smith at home?' A voice asks from outside—it is a middle-aged man, standing at their gate.

'Go, look, Nash, could be the Company messenger. I hope all's well at the mines.'

Nash returns a while later.

'Who is it?'

'Not from the Company, somebody called Remi or maybe Jamie.'

'Rennie? You mean Engineer Rennie?'

'Yes, perhaps. Is working at Summerton at the moment, something to do with digging or some such thing.'

Smith charges out like a mad man. Nash is stupefied. What's taken over his boss? Is the visitor very rich or an old friend? Unlikely. The surveyor meanwhile is at a complete loss about where to seat his visitor. He lugs out his only possession, an old battered chair. And then suddenly turns to Nash and barks, 'Go you fool, get some coffee for Sir.'

Oh! Sir! A Sir to his surveyor sir! A Company man after all then. But Nash has not seen him earlier.

Even as he heads to the kitchen, Nash keeps turning back to look at their visitor. The man is studying Smith's drawings with complete attention, peppering him with questions all the while. Smith never ever shares his drawing book with anyone other than Nash. But this man is happily turning the pages and looking through it. He is delighted with what he sees.

Nash dawdles around, very uncomfortable, and delays putting the water to boil. His mind is roiling. Where has this damned man sprung up from and wedged himself between the two of them? Nash keeps miles away from such natty suited people. Even the surveyor dislikes them. But for some reason he's falling all over himself now.

Nash is deliberately not putting the water to boil. But he's parked in front of the stove. Finally, Smith walks in, 'Why is it taking so long? Isn't the coffee ready yet?'

'We've run out of coffee.'

'Run out, oh dear, what should we do now?' Smith looks across the basin as he says this, and instantly spots a bottle.

'Look there. There's the bottle. You should look around more carefully. Hurry up now!'

In a few moments Nash notices that the man has now stood up to shake the surveyor's hand. 'It was a pleasure, Mr Smith, you will not be inconvenienced at all. Just send back the signed paper please. I'll take your leave now.'

Immediately, Nash thrusts the coffee at him.

'Thank you. I am sorry but I must rush now. Thanks again, Mr Smith.'

The surveyor is looking joyous. Nash grumbles as soon as they are alone. 'Have your coffee, sir.'

'Yes, yes.'

'It's cold now. I'll make one more.'

The surveyor is oblivious to Nash's behaviour. He seems to have had a glimpse of heaven, as though his heavenly messenger has walked out through that very door.

'Who was that, sir?'

'A great man. The Engineer Rennie. Almost all that you see, from London to Somerset, from bridges, tunnels, lighthouses, to castles, are his creation.'

'And why was he here?'

'He's building a canal to transport coal to London from several villages around here.'

'Had he come to tell you of it?'

'Yes, exactly. He wants me to carry out the survey for it. He can dig the canal only after he knows what route it could follow.'

'But you will surely refuse. After all, you are a miner.'

'No, silly, how am I a miner? My work is to survey as the need dictates. Rivers, canals, estates, coal mines.'

'But did you not mention that you had written to the lady owner of our company to give you a partnership in the mines?'

'It's been four months since then. I have received no answer.'

'It'll come, surely. Just give it time. She will reply.'

Smith shakes his head, 'No, no, she will not. Hadn't I told you that I long to see how deep these layers lie underground. If it's a large canal, I'll be able to see so much, so wide and deep. Can you even imagine it? No more down a dark hole, no more down the slippery ladder. Instead, I'll gaze at the expanse of a huge part of England, sliced open with a knife and parted in two just for me. I can't believe my luck, Nash, how can I possibly let such an opportunity slip away?'

'So, you've made up your mind, is it?'

'I'll meet Lady Jones tomorrow morning. Come with me, I say.'

'I can never ever leave this coal mine, sir.'

'Then don't. One must stay where one's heart belongs.'

Nash had hoped that the surveyor would at least entreat him one more time, but Smith is already headed back to his room, his book and pencil in hand.

'I just have a couple of clothes and my book and pencil. It'll fit into a bag. What say?'

'You're leaving, so suddenly, what will our labour say? And what if the lady does not let you go?'

'Why should she not? She's not employed me permanently? She pays me daily wages,' Smith goes back into his room again.

Nash looks up at the sun. The whole house seems drowsy now in the weak sunlight, and yet the roses in the garden are bathed in a warm golden glow. Deep dark red, half blooms. Oliva has seen these roses. She'd asked him for one. Nash had shooed her away. But these roses have no worth for Smith. He'd remarked when his eyes fell upon them at first, 'Why are they so red? A paler shade would have looked prettier.' Nash had looked around frantically and found a white rose plant. It hasn't bloomed yet. The pale white rose right next to the deep red ones. But Smith would surely be overjoyed had they bloomed.

Smith comes out again. He is wearing his pale green shirt. He puts it on every time he visits the pub. He is the same as Nash. They possess no more than three shirts each. Smith sounds elated, 'Come, let's go. Order anything you wish today. If it's vintage for you, I'll be happy to order it. Let this evening be the happiest ever.'

'Are you leaving tomorrow?'

'Why delay when I must leave, but I have loved this house, Nash. A lonely house, just like I am. Everything in it is like me. Even that red rose, as it stands alone. Look.'

'You never liked red roses.'

'I dislike it's deep red. But a single red stalk would look quite fine. Don't you think?'

'Oh!' On any other day Nash would have been overjoyed at this statement. Today, this poetry annoys him no end.

'Listen. Give the rose to that girl.'

'Which girl?'

'The quarrelsome one you spoke of, Oliva?'

'I'll do no such thing. I'll cut the bush down once you leave.'

'Don't be thick-headed. Isn't it great that you've helped a rose bloom in this arid place? If not me, then someone else will come by for its beauty.'

Is man cruel or are his dreams so? Does this man love no one? To him people are no more than pebbles. He feels fire only for rocks and stones. Maybe he hopes for something great, much greater. His dreams have no place for people, no place for affection. This man flows on, unstoppable. Even a river leaves itself behind, under its rocks, but not he.

Smith sits down, suddenly. They love to sit by a gnarled cedar tree on a wooden bench under it. They both love to watch the setting sun from the bench on the way to the pub. Smith is not looking at the sunset today. He is sitting here only by force of habit. The canal will soon be built, in land cut open from this end of the country right across to the other. He will look at rocks, the red, and the green, and the many others. How deep in they go and how wide they spread. The search that has given him sleepless nights, he will now begin that same search afresh. Are the layers he's seen in the depths of the mine spread across England? Or are there others, new ones? Oceans once and land again. What has this country witnessed over the years? Did Noah's great flood come again, and then again? Smith has not read about the earth's creation, not anywhere other than in the Old Testament. Could God have created this earth with a single rock? Why then did he need to create these, layers after layers? Were no new rocks created in these four thousand years? Does it take longer then to create rocks?

One after the other questions come rushing on. They appear, but it's almost impossible to find the answers. And yet the answers are lurking around somewhere. Smith's whole being is trembling with anxiety, and yet he's not at all curious about the person sitting right next to him at the moment. He should have been curious. After all, it was on this bench every day on the way to the pub that they both sat and Nash would belt out endless songs. But today, the boy has not uttered a single word. He is simply staring helplessly at the sun as it sets.

YEAR 1820, BATH

'I don't understand your purpose clearly, Dr Fitton. Why are you here in Bath looking for Smith's fossil collection? I don't have it.'

'There's no need to be agitated, Mr Palmer. I am aware that Smith's collection lies in the British Museum. I will visit it shortly. But I need to find out a little more about the man before I make that trip.'

'Do forgive me, Dr Fitton, but are you investigating Smith's disappearance?'

Fitton laughs, 'Cartographer Cary, Smith's dear friend, had asked me exactly the same. I'll tell you what I'd told him. I'm not a detective. I'm a man of science. I simply wish to know more about Mr Smith.'

'But why? I fail to understand. A naïve, mad vagabond, imprisoned by our government for trying to get his due. A man who had nothing to hide. Why would you be curious about one such as him? Does he owe you, too?'

'No, Mr Palmer, the government is not interested in him at all. I've never ever met him, but I am certainly keenly interested in the man.'

'Influential men such as you are suddenly overtaken by whims. When it pleases you, you search heaven and earth to locate someone. And when not, you don't think twice about setting that person aside, forever. Forget it, Dr Fitton, there is no point in reopening the Smith issue.'

'I have noticed something, Mr Palmer. Those of you, most intimate with Smith, are fuelled by your agitation and anger, and yet none of you stood by him in his bad days. Am I wrong, Mr Palmer? When the man was arrested for his unpaid debts, why did none of you join together to protest the injustice?'

Samborne Palmer is silent for a few moments. Then he says, 'Please sit down, Dr Fitton. I suppose I must miss work today. Please wait while I put the kettle on.'

Palmer comes back soon, an old newspaper in hand. 'We had protested, Dr Fitton. Each one of us had, each one of us who had known him. But not as a group, that's true enough. The man is truly deranged. He did not come to any of us for help. He sent no news. He never informed us of all that had transpired. In truth, he had not understood what being a friend meant. He was clueless about walking the path with others.'

'But you were his friends, weren't you?'

'As I said, he knew no friends, he always walked alone. We watched him from afar. We did try to get close to him, but he was so involved with his own self, he wasn't aware of any one of us.'

'Was he an arrogant man?'

'Sometimes it would seem, yes. And at times he appeared completely insane.'

'And now, what does he seem like now, Mr Palmer?'

Mr Palmer put back his teacup. 'He was like the wind, setting his own course, changing it when he came across obstacles. Have you noticed dry leaves flow along with the wind? It always seems as if these leaves will run along forever. Nonetheless, they do fall upon the earth again. But the wind does not turn back for them. We were the leaves.'

Fitton sips his tea silently.

Palmer stared straight at Fitton. 'Have they begun looking for him?'

'Not officially, no. The mad chatter around him has not yet begun. Ninety-nine out of a hundred people in England have no clue about what that man has achieved.'

'Why then are you looking for him? It doesn't appear as if anyone really cares whether he rots or dies or works or returns.'

'I wish to think on from where your thinking ends. It may be possible to make a lone effort to look for him, perhaps even to locate him. But what next? Will his talent waste away again after that, bit by bit, and he remain just an unknown name like before? It is far more urgent to establish the significance of his contribution than it is to find him. Could I make myself clear, Mr Palmer?'

'I understand, but I cannot believe it, Dr Fitton. He has always been played dirty and deprived of his rights at every step of the way. He is a sad simpleton who cared about no one and yet loved his country intensely. He was invited innumerable times from outside our shores and he had rejected them all. There are men like you and me who revel in self-glory, and then there is he, who refused all such moments of acclaim, silently, quietly, simply for the glory of this land. And that country punished him for his debts, cheated him of his honour.'

'I am aware, Mr Palmer. I have collected enough information before I came to meet you. I will collect more. We cannot give up. I'll approach the authorities of the British Museum and then the Parliament. But not in a state of panic. To appear anxious before the British race is to expose one's own weakness. Those who rule the world are prepared to turn their hearts to stone. We wear masks of cruelty and keep crushing our anxieties within. I need to be prepared to confront this, please help me.'

'Fine. What is it that you wish to know?'

'Have you seen Mr Smith's fossil collection?'

'Some, yes. The entire collection is now housed at the British Museum.'

'Was it his hobby to collect these fossils?'

'To find fossils is a great pastime for the affluent upper classes in this country. It is those who are bored living off their ancestral wealth, that go looking for fossils. They feel entitled to spend their untold wealth on travelling from one place to another to pick up these variedly-shaped objects and decorate their living room shelves with such polished specimens. You know, almost like hunting down the wild and mounting their cured heads in their homes. All you need to do is open any newspaper and find private fossil shows and exhibitions being advertised for most Sundays.'

'Mr Palmer, those you speak of are part of the eighty per cent of our populace. I speak of the rest twenty. For example, the little twelve-year-old Mary Anning who spotted the sea creature Ichthyosaur fossil. The girl's father is an ordinary carpenter who builds showcases for your fancy fossil collectors. And that little child discovered such an amazing creature.'

'There's also a woman named Jeanette.'

'I seem to have heard that name. Was this name listed in the *Dictionary of National Biography*?'

'Indeed, Miss Jeanette is younger to Smith by several years. She gifted a unique piece to Smith once, which too should be part of the British Museum collection.'

'Well, this is what I wanted to speak with you about. Tell me what you can about Smith's fossil collection.'

'What can I say? Just that he picked them up, cleaned and polished them, and sold them all to the Museum. And this marked Smith's end.'

'What do you mean?'

'The British Museum considered Smith to be just like any other fossil collector. Those that collect, decorate and sell off their fossils when they fall upon bad times. No one wanted to find out about the mysteries that he unravelled through these creatures immortalised in rocks. And why do you think that was? Because, if truth were out, the entire nation would be in turmoil. Men's faith would be at stake. It would prove that our religious treatises are not infallible.'

'What could have shaken our religion to its core?'

'There is no force, Dr Fitton, that can hurt religion in any way. Because faith is a matter of the human mind and what it believes in. Truth has its own independent form. But man's imagination forever attempts to change that. We pull one end, push at the other, inflate some parts, deflate others, and keep paring away at it to give it a completely changed shape. Men claim ownership to what they stole from nature in the first place. And those that are fools begin to believe, while the clever use it to their own end.'

'This sounds delightful. Smith sounds like a present-day Galileo or Copernicus. Could this be true?'

Palmer stares at Fitton for a very long time. Then he laughs a strange laugh. It is drizzling outside. Palmer parts the curtains with a long stick. His profile lights up against the bright blazing lamp by his side. Then he says, 'Should I turn off the light, Dr Fitton? It's too bright.'

But he doesn't wait for an answer.

Fitton doesn't try to figure out Palmer's expressions any longer. Instead, he just leans back in his chair.

'It was 1796. I'd just joined the Somerset Canal Company as a sub-engineer. We were told of someone called Smith who had just completed the survey of the canal. This man was to be taken to the other canals during the summer months. We were to inspect every possible canal along the nine hundred mile path from England to Wales. I was taken aback at first. Mr Smith was a surveyor, not an engineer. What would he gain by looking at the canals that were being constructed! So I walked up to the Chief Engineer, Mr Rennie. And what I heard was enough to put my head in a tizzy. The man, it seemed, had no education whatsoever to back him up. Was I expected to be an assistant to an unlettered person? Was I to give up on my self-respect and dignity simply because I had to keep my job? And I hadn't even met the man yet. Mr Rennie's response was, "Fine. Why don't you go and meet him now?"'

'And did you go?'

Palmer perhaps smiles at this.

'How would I? He had three houses, and one was never sure where he would be at any given time?'

'Three homes! I heard that Smith was alone and unmarried at this time.'

'Indeed, he was. The man had taken up residence atop a mound of golden rocks at Tucking Mill on a mortgage. But because the horse ride took up too much time, he'd rented a temporary room at the Swan Inn as well, close to where the new drains are located. He had a third house, mortgaged again, in Bath. No one could ever be sure of where to find him. In fact, I felt more annoyed than astonished by all that I heard. I decided that I would introduce myself only on the day I was to start work.'

'And then?'

'I can never forget that first day. He simply looked at us and smiled. Didn't even bother to ask us our names.'

'Interesting.'

'There's something else I remember about that day. Mr Richard Parkins and I went and sat at our designated seats in the horse carriage. We didn't show him any respect and simply shut the carriage door.'

'Wasn't Smith angry?'

'Forget being angry, he didn't utter a single word. He just fixed his cap and climbed in next to the carriage driver.'

'I can't believe my ears. A Chief Surveyor, sitting with a coachman.'

'And this as the carriage covered miles upon miles under a sizzling sun that baked whatever it presided over. I peeked out of the window to see the man engrossed in the scenes around, scribbling continuously in his notebook. I was forced to ask Parkins about how the man could be completely oblivious to the heat. Parkins believed that the lower classes were used to this, hence…It seems he was used to sitting by the furnace since his father was a blacksmith. I for one liked to talk of specific subjects, usually my estate, as we drove around—about how to transport coal cheaply from the mines, and such like. Smith was still taking in the sights around.'

'What was he looking at?'

'Can't say. The rugged mountains and rolling hills never failed to captivate him. He'd run to the rivers if he saw any there. He always had a small hammer clenched in his hands. He would jump off and begin to tap and strike the rocks around. He could walk miles on miles. We'd be irritated, just waiting for him. We would grumble about our misfortune of having ended up with such a country bumpkin. He wasn't at all concerned about what anybody thought. If he was offered food, he'd eat, if not, he wouldn't. But one thing was obvious.'

'What?'

'Even though annoying, the man was unique. He possessed something which was beyond our grasp. And perhaps because he was so beyond our reach, we felt uncomfortable.'

'Is it because he went around checking things rather than check out the canals?'

'More than that! He'd walk from one end of the canal to the other where a tunnel was being dug in Kingsworth. That's when we realised he was insane.'

'Insane?'

'Absolutely! Work was happening inside that dark tunnel. Each one of us had a helmet with its headlights on, and a gas lamp in our hands. Parkins gestured to us. We went and stood behind Smith. He wasn't aware of our presence at all. He was mumbling away to himself.'

'There are many who do that.'

'No, listen to this carefully. Do you know what he was saying? He was staring at a layer of sandstone and saying, "I didn't expect you to be here. So, the one I saw at the hills of Saperton, was that your brother? No, no, you keep reappearing and confuse me time and again. This is very troubling, how you brought along a new family member. Tell me what you want? I'm quite fed up."'

'Oh my! Insane indeed.'

Palmer's voice drops down to a decibel, 'He would collect a huge number of books. He'd keep turning their pages, whenever he had a spot of time, although he wouldn't read them.'

'He'd look through them, but not read them?'

'Yes, he was looking for something. If asked, he'd reply, "There is indeed something, but I can't tell you what yet. Maybe one day."'

'But that day hasn't arrived yet.'

'You love to talk more than to listen, am I correct, Dr Fitton?'

'Sorry. In truth, I'm getting impatient.'

'In which case there is no point listening. Make up the story on your own.'

'I'm sorry again, Mr Palmer. Please continue. I won't interrupt you again.'

'I can still see that day clearly. I am not a writer, so I may not be able to express myself well. But you, a scientist, I'm sure, will infer much more. A journey that took two months across nine hundred miles from England to Wales on horseback and on foot. And Smith next to the coachman, all the while. A man who showed no emotions, unbothered about his attire, with no need for music or entertainment, with no knowledge of art. A man who would make excuses to be excused if invited to a party. One cannot spend a moment with such a man if it is not for work. A man who would collect fossils all day long and stay cooped up in his room cleaning and polishing them. A man one cannot call normal.'

'Completely unsociable. Oh, dear. I've interjected again.'

Palmer doesn't say anything. He seems lost, his mind far away.

'That day we were sitting around like tourists at the tower of York Minster. We'd taken around two hundred and fifty steps up to the top. We were all panting from the effort and sat down to catch our breath. One could see clearly all around, a clear view of thirty miles. Smith seemed to be in wonderland. He said, "Do you notice the mountains, Mr Palmer?" When I said I do, Smith continued, "Although I can't be definite, the contour pattern and sparse tree growth suggest these are chalk mountains." I said, "Wonderful, truly Mr Smith, your powers of observation must be lauded." Parkins and I looked at each other and winked, for who could tell about the rock formation from this far. Smith answered, "That's not how I meant to say it, Mr Palmer. There is a pattern here which you are not able to understand. Would you remember the layers that you had seen at the canal of Kingsworth in Birmingham?" "Clearly. Coal with sandstone on top and then red marl above. Three distinct layers, one on top of the other." "The sandstone just about resting on the coal." "What are you getting at, Mr Smith?" "Give me your left hand, Mr Palmer." As soon as I did so, Smith sandwiched my hand in both his. "What's this, Mr Smith?" "Just consider this, Mr Palmer. The three layers did not form at the same time. The layer of coal was formed first. It's the oldest rock strata. Sandstone overlies this coal, and the youngest rock layer right on top is red marl. The oldest would naturally be at the bottom and the youngest on top." "I am afraid, I do not understand, Mr Smith." "All stones and rocks age just as we humans do. They too are young and old." "But the Bible states that God created it all in one day." "Mr Palmer, you are an educated man. Use your logic to understand this. It takes millions of years for sediments to be deposited in layers and these layers to convert into hard sedimentary rocks. If only we had an instrument to measure the age of these rocks!" I asked, "Is then coal the most aged?" He answered, "I do not know because I have not been able to descend to the deepest parts of the earth. If I were to say that the granite we saw on the mountain top this morning are even older, would you believe me?" "How do you know?" "I'll explain. Forget everything except our hands. My left arm is at the bottom, then your left arm on it and right on top, my right. Now these tilt to the east. This means, the further east you move, these layers will sink deeper into the ground. On the west, you'll see these high above the ground. First you will see the layer at the top, that is the tip of my right hand, and further west, you will see your left hand and then, my left hand, which is the bottommost one. Now we see red marl along the canal walks and then sandstone above the ground. Had we moved further, we'd see coal. We'll test this tomorrow morning.

Further west from these, I bet we'll see granite. Should we test it tomorrow, Mr Palmer?"'

'Fantastic. He established the law of superposition of sedimentary strata. Did you explore it further, Mr Palmer?'

'We'll come to that in a short while, Dr Fitton. Hear me out on what I experienced at the moment. In the falling light of that evening, Smith held my hand in both his. I will never forget that touch. It was as if Copernicus had come alive and was admonishing me, "Have you not understood yet how the sun holds still and the Earth revolves around it? How long can you ignore the infallible truth with your made-up tales?" Dr Fitton, do you hear me?'

There is silence. Almost as if the man who is always interrupting and offering explanations, has now fallen under a strange spell.

Palmer offers, 'If you still have any doubts, let me assure you that the very next day as we walked on and on, we came upon red marl, then sandstone, and where we ended our journey for that day, just ahead by a few kilometres began layers of coal. We were dumbfounded, both Parkins and I. A new door had swung open in the annals of science. There was so much work happening on the present times, but that it was possible for one to look into the geological ages, into the formation of the earth millions and millions of years ago, I saw clearly for the very first time. That man, a man without a degree, a man without a lineage attached to his name, a weird, strange man, had understood this so well, the play of the creation and destruction of our very earth. Parkins had said at that very moment, "I pray to God that this man continues his work for another hundred years." I, too, had called upon the Almighty. But my voice had probably gone unheard.'

YEAR 1799, BATH

Bath, a city at the crossroads of the modern and the pre-Roman eras. It is believed that Minerva, the Goddess of Wisdom, would bathe in the waters of its warm pools. In the later years, many huge bathing houses were built here. Now, tourists from across the world, come here for a dip in its thermal pools.

London is a city that dazzles. With its parties, clubs, theatres, and musical entertainment. It's a city that captivates everyone. People of refined taste do not wish to leave it at any cost. Bath is dull in comparison. Bath has thus developed a distinct identity as a town that eschews the hustle of festivities. It has become a seat for the pursuit of science and philosophy. The first man to pursue scientific research in England took birth in Bath. Adelard of Bath. It was he who first formulated the idea of a scientific consciousness in the 12th century. Although largely forgotten today, at the end of this 18th century, the city's scientific community continues to hold a position of pride.

On 22 December 1796, Bath witnessed a meeting. Exactly nineteen years earlier, in 1777, a certain Society had been constituted here—the West England Society. Several rare geniuses had participated in the first meet—Joseph Priestley, the discoverer of oxygen and other gases, the economist Thomas Robert Malthus, the renowned astronomer William Herschel who discovered the planet Uranus, as also Humphry Davy, known for his discovery of the elements sodium and potassium among others. Since then, the Society has seen nineteen such annual gatherings. These luminaries brighten the meets with their brilliance.

On 22nd December, the Society, celebrating its twentieth anniversary, made two announcements. About the first announcement, the worthies had already been informed. But the second announcement set off a ripple of murmurs and whispers amongst its members. What has come upon the President of the Royal Bath Society? They've only just completed twenty years of age and he has muddied the Society with ditch-water. What else can one call it? He has had a declaration made to confer the Fellowship of the Royal Bath and the West England Societies on members such as the Duke of Bedford and the Earls of Peterborough and Egremont. But, in the same breath, he has also announced the inclusion of a nameless villager who has no lineage to align to! This is the offspring of a blacksmith, without an iota of social grace, a man without a clue to art and theatre, who knows nothing about parties or clubs. He has not an estate or a building to his name. How dare a man such as him aspire to the Society?

The man's young, too. Burnt by the sun with only a shirt to his name, one can easily see how low he looks, almost like a labourer. It seems he's renowned for his surveys on England's rivers and canals. But there are countless like him everywhere.

A round of food follows the meet. The older members of the Royal Bath are congratulating the new entrants. There are bouquets aplenty. But not one man there finds it to his taste to greet their fourth new member. The man in question is unconcerned. He is devouring his food in silence. Almost as if he knew how this would unfold.

'Mr Smith?'

The man looks up from his plate. A large man, smartly turned out, is staring at him.

'May I sit with you?'

Smith nods. The gentleman takes the seat next to Smith.

'I am Reverend Richardson. I am a priest, although I carry out a special duty outside of my religious practice, a hobby you could call it. It makes me very happy.'

'I see,' Smith continues with his food.

'I have a request, Mr Smith. I've collected numerous fossils from across the land. I've prepared four chests of drawers for it.

But I am unable to arrange these according to their geological ages. What I mean is, I do not know which of these are older and which younger.'

Smith says, even as he continues to eat, 'How would their age matter? After all they were all created in a day, or weren't they?'

'According to new information trickling in, that's not how it all happened.'

'How would you, a man of religion, go against those beliefs?'

'That there is a clash between logic and religion, I do not believe, Mr Smith. When the extent of men's knowledge comes to a halt, other tales begin to take shape. The paths of knowledge open up again when men like you arrive to clear out superstitious beliefs. Men begin to walk on those new paths again. Religion is an open expanse like the skies. Our path of logic is like the sky, too.'

'Is that true?' Smith wipes his mouth with a napkin, his food done.

Richardson maintains his natural calm. He is Rector of the church. He comes in contact with an array of people. He is used to imparting the essence of religious beliefs to hardened criminals as well. He tries to infuse them with well-being. This man in front is far more respectable than them all. He cannot judge the man yet.

'Mr Smith, would you come take a look at my fossils in my terrace house?'

'I will.'

'If not a problem, would you arrange them methodically according to their age?'

'Sure.'

'When, Mr Smith?'

'Tomorrow.'

'Tomorrow!'

'I won't, if that's a problem for you.'

'Not a problem at all, I'll be honoured.'

'Good night, then.'

The man leaves the party. Reverend Townsend comes and stands by Richardson, 'Now where is that incomparable being of yours?'

'He left a moment ago.'

'Without a goodbye to any one of us? Teach him decorum. He's been made member of such a prestigious body, not many are blessed with his good fortune.'

'I doubt he understands concepts such as respect and honour. I'll see what I can do when I meet him tomorrow.'

The night after, there is a knock on Townsend's door. The Reverend has just completed his dinner and is readying to be at prayer. A knock at this time of the night means someone's in trouble. He gapes as he opens the door.

'Richardson, what's the matter?'

'Let me enter, Father. It's cold outside. My feet feel frozen already.'

'Come sit by the fireside.'

Richardson is barely able to speak. His face is brushed loosely with snow. He grabs the coffee as Townsend offers it to him.

It takes him about half an hour to warm up enough to speak again. The first word he utters is, 'Unbelievable!'

Townsend is an intelligent man, in fact his mind is as sharp as nails. He knows immediately that there is no trouble. It's something that has Richardson all excited. He waits. More will be revealed soon.

'Mr Smith arranged everything in a day, Father. He knows by heart, every little detail about the Bath fossils. What came first and why, what next, and so on, in great detail. He is a genius. The world needs to know him. We must do something about it.'

Townsend smiles faintly, 'Curb your excitement, Father Richardson, it doesn't suit those who work for the church.'

'I know I am excited. But how can I disbelieve what I saw today with my own eyes?'

Townsend does not say anything in response. It is obvious Richardson will take time to calm down. He can wait.

Richardson asks again, his voice laced with disappointment now, 'Father, will you not do anything for the man?'

'What can I do? The one who can, will.' And he touches the cross he is wearing round his neck.

'Indeed, He is the one who does, but are we not the medium? The upper classes of England will not let him rise, this simple, naive man, one who knows nothing of social complexities, its twists and turns. He continues to work silently, holding his birth responsible for all his misfortune. No one understands this, perhaps no one ever will.'

'It seems you have understood him well in just a day. Such emotion is uncalled for, Richardson.'

'I haven't understood him much. Only this, that if we ignore him, the progress of the world is doomed.'

'That's a daring statement to make. The world will not progress without William Smith! Has he completely mesmerised you?'

'He knows nothing of magic, Father, he has never been taught the ways of the world. All day today, he spoke only two sentences at most. He asked to be shown the fossils, and then while leaving, "My work is done."'

'How then did he explain about the ages of the fossils?'

'He didn't. He wrote it all down. Father, were you not a doctor before you took on this frock? I, too, was a student of the natural sciences. If we, too, sit tight, unmoving, like all the others, will not our future generations call us out?'

Townsend lets out a deep sigh. 'I understand. Bring him here one day. I need to see this clearly for myself.'

'Bring him? When?'

'Sometime next week.'

But the next week Smith was nowhere to be found. He had already left for Paulton Halt on work for the Canal Coal Company. He would not be back for the next two months. By which time Reverend Townsend would have left the country for six months. By the time he returned, Smith had left for Limpley Stoke for some other work. Father Richardson, by then swamped with work for the church, was not always able to keep a tab on Smith's comings and goings. It was only two and a half years later that he managed to track Smith down. Although not unemployed, Smith was very close it.

Richardson had to almost abduct Smith to get him to Reverend Townsend's home.

Townsend, the adept man that he was, meanders through a long conversation about this and that with Smith. Suddenly, he asks, 'Do you still work for the Canal Coal Company, Mr Smith?'

Smith, unruffled, answers, 'Not anymore. They've thrown me out.'

'When did this happen?'

'In the month of June.'

'And why, may I enquire?'

'They asked me to plan the canal as was convenient to them, to use cheaper material for more profits. I do not work like this.'

'There is no other reason, nothing personal I hope, Mr Smith?'

Richardson figures out that Townend has followed Smith's journey very closely. In fact, everyone in Bath knows of the reason Canal Coal has presented for their dismissal of Smith. But Smith himself is unable to understand what Townsend is getting at.

'There was no personal interest,' Smith answers, 'I had rented a house from the Company. The Company wanted to take it over, but I told them clearly that it wouldn't be possible for me to work if they did so.'

Richardson interjects, 'They haven't given Mr Smith his pension either. Not even a compensation for dismissing him from service without notice.'

'They didn't even say goodbye to me.' Smith's sentiments are clearly hurt.

'That is immaterial. But a pension for such a long service is a must. How could it be withheld? You should have filed a case, Mr Smith,' Richardson says in a grave voice.

Smith seems to think of something. Then he purses his lips and says, 'Pension be damned. I've been saved from working with those crooks.'

'Are you working elsewhere, Mr Smith?' Townsend, who has been sprawled across a chair all the while, sits up straight. It's urgent now that they discuss things seriously.

'I am free at the moment. I go and provide my services whenever asked. Right now, I am measuring a mine.'

'That's excellent. I hope you're able to earn something.'

'Yes, indeed. And, in addition, I've been able to travel across England.'

'And so your life will go on thus, is that what you've decided?'

'In what way?'

'I mean, you will serve whenever you are called. They'll hand you a few pounds, and then they'll throw you out. And then, you'll proceed to wherever you are called next.'

'It's quite alright, truly.'

'You are afraid to dream, isn't that true, Mr Smith? Those who dream are masters of the dreams they wish for themselves. You, I think, Mr Smith, were never taught to dream.'

Smith continues to stand, stricken. He does not know how to answer questions which are not straight. What is this man asking?

Townsend continues, 'Have you heard anything about me, Mr Smith?'

Smith answers gruffly. 'You left the country in fear of a woman and have only just returned after some years.'

Taken back, Father Richardson is about to interrupt Smith when Townsend gestures that he shouldn't.

'You are correct, Mr Smith. The lady in question is evangelist Selina Hastings, a woman from an affluent family. She has helped build several churches around Bath, not only in keeping with the tradition of her family, but also in her individual capacity. I was a personal Rector to them, amongst a few others. She asked that I not give any sermons or talks outside her church. I ignored her injunction. Every human being has the right to participate in these sermons. Why should a person be deprived of this even if they are not members of a church for reasons best known to them? I thus earned Selina's wrath. She and her pet ruffians made my life hell. So I had to leave the country. Now do you understand what I'm getting at?'

Smith nods silently.

Townsend laughs silently, his eyes brighter than before. He walks up to the statue of Christ impaled on a cross.

'I did not run away, Mr Smith. And I have returned. Do you know why? Because I used to dream of how I'd take religion to the common man, telling him of it in words easy to understand. And of how I would also awaken their scientific spirits. There is no need to hide behind stories to know the plans of nature. We cannot suppress man's curiosity by filling them up with a fear of religion. If we, preachers of religion, do not come forward, they will never lose their fear.'

Smith is lost in thought. Townsend takes his hands in his. 'What are you contemplating, Mr Smith?'

'How beautifully you speak, Father. I had not realised how magic fills up your voice and words. Father Richardson was right about you.'

'What has he said? He loves me too much.'

'You deserve the love, Father. The power of God seemed to flow through you as you stood by his form and spoke. Your voice was so incredible, I'd lost my ability to move listening to you.'

Townsend gently presses Smith's hand. 'You must share with the world what you have seen and felt today, Mr Smith. You are a scientist. We have just begun to understand science. How can we now afford to sit quietly? Do we not need to walk that path? There are many who earn wealth, Mr Smith. How many can vouch to have earned knowledge? Walk on, Mr Smith, walk ahead on your path.'

'What must I do, Reverend Townsend?' Smith's voice is choking.

'Do I have to tell you that as well? Listen then. You must draw geological maps. But not like the ones for the mines. This map must show what lies below our earth just as you would show what can be seen above.'

'I had attempted one, Father, but I failed. This can be shown as a cross section, but it is perhaps impossible to draw a three dimensional layer on a two dimensional map.'

'It will be possible, and you, only you, will make it so. Think hard, Mr Smith. You are the master of all three subjects—surveying, engineering and geology. To hold this up to the world, you must take to pen and paper. I invite you to dinner on this day three months hence. The three of us will be present. Then we will present it to the world. Dream on, Mr Smith, we are with you.'

***

Three months later, when all of Bath is in the deepest of slumbers, Townsend's Pulteney Street house is still lit up. On the table is spread William Smith's map. Everyone is familiar with Taylor and Meyler's geographical map of Bath. On that map now are shown the details of rocks right from the top of Dundry Hill to its westward tilt all across, with three layers of chalk, oolitic golden limestone and coral limestone. In the last seven days, Townsend and Richardson have walked over every inch of Dundry, Smith's map clutched in their hands. They have dug holes across the expanse to make sure about the details of the newly drawn map. They have looked at each peak and nook, small and large, on the mountain. The map cannot be faulted at all.

Richardson takes a seat in front of a candle. The three layers of rocks can be seen in three different colours, chalk a bluish white, the golden limestone a distinct yellow and the layer of coral limestone in red. Smith is speaking and Richardson is taking notes. Once they're done, two more copies will be made. They will work into dawn. Reverend Townsend has only one task today. To make coffee, and wash the mugs over and over again, through the night. After serving coffee again at two that night, he says, 'Enough. You'll feel ill if you drink this anymore.' The same man says again at three, 'Perhaps I'll make a light cup for us again. Let me light up the stove once more, yes, let me do that. But it's a pain to light this stove.'

Richardson looks up. This was the man who was mumbling, 'pain, what a pain,' all the while. Now his eyes are shining bright, just like the morning light. Man is truly a strange creature.

YEAR 1820, CHARLCOMBE

Monday. The morning service of the church has just got over. It's about 11:00 right now. It's not Easter, not even a holiday. Just a normal busy cloudy day. But the church is overflowing. In these small suburban areas, no one is ever in a rush. Many flock to the church, therefore, to listen to stories about God. And these days, the school exams are on. So many students come to the church without fail. Those who attend the service know each other well. It is a small town after all. Father Richardson notices an unknown face sitting silently on the left of the third bench. He's not too surprised. But he has made a mental note of the fact that the man is not a local, nor is he a guest of someone from the village. The matter becomes clearer when Joseph, the church pianist, comes to inform him that someone is waiting outside to meet Richardson. The bell for lunch will go off at 12. It's 11:20 right now. Since the man has not taken an appointment, Richardson can easily refuse to meet him. But he hesitates to do so. The man is an outsider.

Why would he wish to see Richardson unless there is a special reason? Richardson will give him about half an hour. If that's not enough time for the man, he'll ask him to come on another day. Richardson comes out. The man's waiting on a bench under a crooked fir tree. Why has he chosen a dark damp corner when there's so much open green around?

Richardson walks up.

'Good morning, Father Richardson.'

'Good morning, may God bless you.'

'Father, I am Brother Robert. I studied for my Rectorship under Reverend Townsend. He's sent me to you.'

'Is the Reverend here in Bath?'

'No, he's been in London these last few months. This is a letter for you.'

'I hope all is well. Why would he send a letter through you?'

'I wouldn't know for sure, but I was to come to Bath for training, perhaps that's why.'

'And do you wish to take back my reply with you?'

'He didn't mention anything about that. But I could wait if you wish. In any case, I am stationed in Bath for a few more days.'

'I'll need to read this first. Why don't you rest a while, Brother Robert? The main town is not too far from here. You could get back there in the evening.'

Richardson hands over the care of Brother Robert to Joseph. He looks over the envelope a few times. The fir tree is perhaps a good place to read the letter in peace. It may contain some private matter. He begins to read:

My dear Father Richardson,

By the grace of our blessed Almighty, I hope your life in the service of the church is moving ahead smoothly. I've come to London for a special assignment. I nurture a secret desire in my heart. I was hoping to meet Smith at his Buckingham Street home soon. But I have heard some news that has left me astounded. I am told Smith has left his home and disappeared. No one knows where he is. Smith had returned to this address after serving his prison sentence. But he had been barred from entering it. The building has now been taken back by the house owner. They have somehow not touched Smith's office. Smith's assistant, Mrs Kitten, sits there oftentimes but not always. For some unknown reason, she's been allowed to continue. The house owner has, no doubt, bought a few of his valuable possessions and a piece or two of museum collectibles to free him of his debt, but he has allowed no one to touch personal papers, maps and collection.

On the other hand, a geological map covering England, Wales and Scotland, priced at 5 guineas, has been published by the Geological Society. The name of the mapmaker mentioned is Bellas Greenough. You will be dumbfounded if you take a look at this. The exact twin of what Smith made five years ago, and yet Smith finds no mention anywhere at all. Not even a single courtesy mention. A close friend of mine has raised the question with the Society—if the maps were placed one beside the other, it would be as clear as light that the details have been stolen. But who is the thief? Smith or the Society? There is nothing further to be said. It is the responsibility of the Society to prove that Smith is a thief.

Almost twenty, perhaps twenty-five years ago, some of us had sat through the night and a map had taken shape in my Pulteney Street house. A report had also been readied. We three each possess a copy of that map, that is you, Smith and I. The streets shown in Greenough's map are a complete copy of that report. I am at a complete loss about how he came upon the original.

Greenough's map is riddled with errors, and yet it will sell well because it is priced low. We must protest strongly and swiftly against this injustice. But how should we go about this? After all, the man at the centre of this matter is missing. A simpleton who knows only to curse his birth and fill himself with regret, but knows not how to claim his due. Would you be able to find him? We must locate his whereabouts immediately.

The sin that was borne by the Son of God, is the sin that man must bear now! I pray to you, holy Son of God, may you free men from committing further sins.

Richardson cannot read further. He shuts his eyes tight and sits silently. What can his reply be? Smith has left London. No one knows for where. He's been missing for a few months now. Those who are acquainted with the matter, have not tried to look for him. Although Smith has many who oppose him, the numbers are not more than his friends and those who love him immeasurably. And yet he is now drowning in a heartbreaking situation, with no hope of relief. Who is responsible for this terrible act? Smith maintains it's his fate and birth. He has been victim to England's hierarchical mindset time and again. But what of his publisher, Cartographer John Cary, who too is a man from the middle classes? His father was a small-time malt maker turning corn into malt. And yet Cary's published maps are to be found in every home in England.

In truth, what man knows as fate is not really such a simple deal. It is a complex mix of birth, upbringing, nature, intellect, activity, society, family, friends, well-wishers, enemies, a coming together of all these very factors which carry one forward, propelled by an invisible force.

Quite like the dough that makes a cake. How well the cake rises, how good it tastes, will be determined by the ingredients that go into its making. By blaming one's fate, man is indeed refusing to acknowledge the force of the ingredients of his life.

Some eighteen years ago on this day, Smith had pulled off an astounding feat. In this very town of Bath, the source of warm underground waters had dried out. There was no water to be found in the city's reservoirs. The mayor was totally at a loss. Bath's importance and earnings depended largely on its thermal water pools. Smith was supposed to be in Bath at that time, carrying out his survey work. But when word was sent to him, they realised he was in London. Word was then sent to London.

Smith dropped all his work and came post-haste to Bath. His verdict was to dig a hole in the ground. A smallish hole that would tunnel down to the bottom of the pool area. It was hard work getting the Duke of Bath to agree to this, to dig a deep hole right next to one of the city's important pools. And also, someone had to climb down the hole to the source of the water. But the closer one went to the source, the higher the heat would be. The temperature would increase and go up close to 150°F. No one but Smith would be capable of carrying out this job. When Smith reached the source, a large ox bone was found embedded at the mouth of the source, blocking the flow of water. Silt and sand had over the years made the bone's structures as hard as stone. But thanks to this naive man, William Smith, the Duke of Bath could once again hold his head high.

And now this, what a terrible fate for a man like Smith! A man who even the Prime Minister of England had wished to meet, of his own, without a request or invitation, in his Buckingham Street house. The man, the first edition of whose map had sold four hundred copies at one go, unbelievably imprisoned, for not being able to pay off a small debt. Today he is an impoverished vagabond, a man on the run, a man who has lost all hope of holding onto his fame and fortune. Smith was prone to a strange lassitude many a times. He had been warned, therefore, against possible negative outcomes.

Once Townsend had said jokingly, 'Since Smith is not doing anything at all about his work, it might not be a bad idea for me to publish a detailed paper on my own fossil collection.' Townsend's statement had been made in jest, but it had made Richardson pause and think deeply. If Townsend could make such jokes, what was there to stop the many others from actually doing so? Smith had then been told to start writing his memoir immediately. How long could he keep his research and facts unused?

Smith hasn't been bothered. And how could he even be? He was after all running around from place to place, drunk on work. Oxfordshire, Yorkshire, Gloucestershire, Hampshire,

Derby, London, Bedford. Where would he even find time to write a report? A brown woollen shirt on his back, an oversized hat on his head, his map folded in his suitcase, he was busy moving from place to place. And if he travelled by carriage, he always sat next to the coachman. His face was gnarled beyond belief, baked in the sun, dampened by the rain, he seemed like a soldier who had been battling at the front. His food—apples and nuts taken from the trees on the way. Onlookers were sure to be shocked—was this a man or a creature from an alien planet?

It is never enough to work like a man possessed. The times demand that effort must be made to let the world know of one's successes. It is no longer the world of thousands of years ago, where men praised each other selflessly and fame spread by word of mouth. Smith's research had begun to get stolen, bit by bit. A writer named Richard Warner published a book on the history of Bath. In it, he detailed the structure of the earth's layers as Smith had deduced, details that were an exact copy of Smith's descriptions of layers, old to new. And yet, no mention of Smith at all. The Society of Arts was asked to justify this matter; no answer was ever received. Everyone was aware, and yet they maintained their silence, for it hurt their ego to honour Smith. Even in this instance, Smith hadn't pressed any charges. Instead, he said, 'How much can they steal? I am an unending store of facts.'

Reverend Townsend had added as PS:

Would you have any recent information about Smith's wife? I'd been told that the woman is incurably ill. Would Smith have taken her with him in such a situation? If only we could gain some information about them from her parents!

Why would Reverend Townsend ask Richardson about Smith's wife? Is he pointing to some particular matter? Richardson folds the letter and shuts his eyes once again. He can see his alter ego so distinctly. He, too, is a preacher named Richardson. He's exhorting loudly.

'Are you saying that you have never meant any harm?'

The Reverend, his eyes still shut, mumbles, 'Not while in my senses, no.'

'An ideal partner can help man attain the doors of heaven, so say our religious texts. To allow an unlettered seventeen-year-old to marry a sensitive youth? Was it at all justifiable?'

'No girl from an upper class educated family would ever agree to be the wife of a blacksmith's son.'

'When he married, he was a reasonably successful, well-known man, with reasonable means to keep himself. There were indeed several offers from affluent and respectable families, were there not?'

'Yes, perhaps. But Smith was not at all concerned about marriage or to start a family. I knew the girl's father, a farmer.

He was on his death bed. He begged that I find someone for his daughter post-haste. Or else, she would be left an orphan very soon. I mentioned this to Smith and introduced him to the girl. Smith agreed. Was I wrong?'

'You defend yourself well. You helped out a dying man, rescued an unlettered girl from an uncertain fate. And your friend agreed to help you out. The wedding was indeed held in the church as all weddings must be. But why the secrecy around this union? What was there to hide? Not one of Smith's friends, and by then he had many well-known social figures as friends, knew of the wedding. Why? Why, indeed?'

'Smith hated crowds, disliked ceremonies.'

'Yes. That no doubt is true. But is it not true that you did not wish the news to spread, you feared disruption, and obstruction to this union? Because there were many who knew that the seventeen-year-old Mary had been mentally unwell for many years since childhood. She suffered a medical condition called nymphomania. And awaited a slow death, that weakened her mind and body, one day at a time. She would never bear a child. A happy, even normal, life would never be possible for a man with a woman like her.'

'How could I abandon a helpless child like her? I had made a promise to her father.'

'Why was marriage the only option? Why not treatment? Why tie a simple work-loving man to such a mad woman?

Smith trusted you, but you denied him even this little family, this happiness. You closed all doors for this poor man.'

'I did not know then, I truly did not know.'

The bell for lunch is ringing away, but Richardson is deaf to it. Joseph is standing by him, calling him, but Richardson is silent, lost. Joseph is wondering what he should do next. Father's eyes are shut tight, and he is mumbling away. His face is wet with tears. Is he praying? Should he even be left alone in the state? Joseph runs back to the church.

YEAR 1807, LONDON

'There is a letter waiting for you, Mr Smith,' Mrs Kitten waddles into Smith's room carrying her bulk quite efficiently.

Smith is neck deep in work. Last week, he'd bumped into a very influential figure while he was out and about on work. As soon as the gentleman saw and recognised Smith, he had hollered, 'Strata Smith, isn't it?'

At this exact point in time, Smith was busy observing a bluish Lias stone, sitting squat next to it. The other man seemed almost ready to pounce on him, so excited he was to spot him. 'I've found you!' His exclamation was as surprised as when one has found an escaped fugitive or a couple that has eloped.

Smith is usually always lost to the world when at work with his beloved stones. Many a times, he's been saved from being rundown. This is exactly why people dislike working with him. He might be headed to survey some mica mines, but will jump off his carriage if he spots an unknown variety of rock. And then, hours on end, he'll be parked near it, staring, observing, measuring, contemplating. He uses his hammer and chisel on it, puts labels and so on. And he would have clean forgotten where he was headed in the first place, already running against the clock. The problem is that unlike all the other surveyors and engineers, he cannot differentiate between what needs to be done and what may not qualify as important and urgent. And, therefore, he takes a lot more time on each work than has been allotted for it. Those who employ him end up irritated, sometimes deducting a large portion of his due. Nowadays, mine owners have become alert to these Smith-related issues. They inform all new clients whenever recommending him, 'The man's work is impeccable, but of the time he might take, nothing can be promised. If you could blindfold him while he is being transported to work, perhaps your work will be done on time.'

The wary clients, of course, enquire in advance, 'Is he working for others at the same time?'

'No, that's never the problem. He will not ever breach any contracts or go against his word. But how he goes about his work can never be detailed, it has to be experienced on one's own.'

The excited man now touches Smith. 'Sir, are you not Strata Smith? William Smith, the blacksmith's boy?'

Under the hot burning sky, a strange man, all nervous energy, wants to know if he is who he is. William is now known more by this new honorific, Strata Smith. But why should his lineage, his birth be brought in to identify him? And loudly! Is this man trying to insult him?

Smith tries to sound grave, 'And you?'

'I? I am a farmer's son. My father produced malt from cereals. I sell atlases, I am John Cary.'

'I seem to have heard of you. There is a publisher John Cary, right?'

'I am the same.'

Smith is wonderstruck. He wipes his dust-coated hands on his dirty coat and extends it for a shake. 'Fancy, yeah, it is you. What are you doing on this forest path, sir?'

'Sir? Who is that? We belong to the same tribe. Didn't I mention that? I am a farmer's boy? And you are absolutely wrong if you think that wandering these forest paths is only your prerogative. You know that I am a surveyor, too, although not as famed as you are.'

'Don't embarrass me. Atlases published by you can be found almost in every home in England. Are you here especially for some work?'

'Yes, yes. Not here in the forest but a place on the main avenue. I was curious when I spotted your carriage. Your coachman mentioned your name. Although, I must say the man looked mighty unhappy. I believe you have to travel another twelve miles or so.'

'Almost, yes. But that shouldn't put him off. He'll be paid for extra hours.'

'And that payment you will make from your own pocket, I suppose? It's exactly as I had heard about you. You are trying to deliver the cause of science at the cost of your own future.'

'Let that be. What are you here for?'

'Do not even ask. The General Post Office has assigned me the task to ready a new atlas. A map that plots the roads of all of England. And this is what they've given me to complete the job. A bicycle!'

'This! Looks like a fine piece.'

'It is, an excellent one. I use it to travel and measure distances, and bends and curves. It's also a great crutch when I am tired.'

'Indeed, exceptional!'

'But let me tell you something. In truth, you are wasting yourself without profit. And although you are discovering the biggest mines across the country, can speak extempore about the patterns of all our rivers and have seen untold wonders across our land, I do not see anything written by you in any of our known journals.'

'This is my true vocation. I know exactly what lies under the soil throughout England's expanse.'

'Is that even possible?'

'That is for you to decide, but I know it.'

'Will you be able to draw it all? Map it all onto atlases?'

'What if I say yes?'

'Are you serious? If this was true, then my atlas would be a historic business. It's beyond my imagination. You are a man who has carried out this tremendous work, silently and single-handedly. By when will you be done?'

'Say another fortnight?'

'Fortnight?' John Cary seems to be running some calculations in his head. 'As for me, I'll be done with the first phase of my assignment in another month. Well, then, drop in at my office on the 12th of next month and let's talk.'

Today is that day. 12th of October.

Smith returned to his Buckingham Street home last night. The mica mine job took almost a month instead of the planned fortnight. But Mr Slater, the mine owner, has paid him wages for two weeks only. He has also expressed his annoyance at the delay. In fact, he's gone on to say that he was lenient because of his compassionate nature. Any other owner would have extracted a fine, too. However, just having said all this, he'd also asked if Smith would like to continue work for his stone quarries.

This incident is not uncommon for Smith. It happens often to him. Reduced wages, sharp words thrown his way, and then an offer for continued work.

Smith is now putting his papers together. For ease of work and transport, Smith often cuts large maps into smaller manageable sizes. Now, first, he must number them correctly. There's only an hour left for him to start for Cary's office.

As Smith looks up at Mrs Kitten entering the room, he bumps against the lamp. On any other day, she would have run to him with ice or cold water or even looked for the first-aid box commiserating about his well-being. But not today. She only leaves a glass of water on his table. Clearly, she is angry.

'I hope I have paid your salary for the month, Mrs Kitten? I'm afraid I'm prone to forgetting it often,' he offers, hoping to find out the reason for her annoyance.

'Do you think this is about my salary?'

'No, not at all.' The thirty-eight-year-old Smith is extremely heedful in his behaviour towards the roly-poly Mrs Kitten. She's a sensitive person, but wonderful and affectionate. And yet she is given to moods, easily angered.

'You brought in a letter, Mrs Kitten. Do read it for me, please.'

'I absolutely cannot. It's a personal letter, posted from Churchill, perhaps. It carries the Oxfordshire stamp.'

'I'll read it later in that case, please give it to me.'

Mrs Kitten continues to stand, ignoring this. Her countenance is still serious. Maybe it is something else.

'Should we speak later, then, Mrs Kitten? I am in a bit of a rush.'

'You will never have the time. You are forever rushing from one to the other place. Who knows, you may return now only after a month. You must therefore hear me out.'

Smith plonks down on a chair, 'Fine then, tell me. I'm all ears. Please pull up that chair.'

It doesn't seem as if Mrs Kitten will listen to him. She continues to stand, still serious, her arms folded across her chest. 'How long do you hope to continue like this, Mr Smith?'

Smith is flabbergasted, 'What exactly do you mean, Mrs Kitten?'

'You've spent all dues to the coachman for the last four weeks from your own pocket. You always spend double of what you ever earn. That is exactly what I mean.'

'Ah! That. In truth, no company ever spends on work that they consider of a personal nature.'

'Personal? Fine, I'll go with that. I am always fighting the odds just trying to keep your office running. You have bought four, maybe five, houses across different places, or are paying mortgage on them. You have kept rooms booked under your name in at least two hotels. May I know why a man, alone and single, needs so many accommodations all at once?'

'Those rooms are not for me, Mrs Kitten, but for those who tag along with me on my shoulders. Wherever I work, heaps of stones and fossils just cling to me and climb all over me, asking to be carried along. I don't have the heart to abandon them.'

'Eight guineas are spent each month on this five-storeyed Buckingham Street house. How does that help you?'

'You must understand the historic significance of this house, Mrs Kitten. I believe Charles Dickens lived here for a while. And then, the Russian Emperor…'

'That's enough, please, stop. They've told you stories and fooled you into this. A man such as you, highly educated, so full of knowledge, and yet so gullible, it's unbelievable.'

'You forget, Mrs Kitten, I have no education, no degree whatsoever. Perhaps that's why they fool me so.'

Mrs Kitten pulls herself back. Perhaps, she's been too harsh. She is always concerned about this sensitive, emotional man. But she must somehow make him see her point. It will not do to step back today. He is frittering his life away. There is no one to stand beside him.

'Oh, fine then. Let's assume you are correct. And this house is indeed extremely valuable and historic. And yet, you have not spent even a full week here in this huge palatial place. Not a single curtain or even a carpet has ever been replaced here. It has never seen a fresh coat of paint. In fact, the house has never ever seen a single dinner being hosted.'

'You are right, Mrs Kitten. And yet, this house has had the presence of the Duke of Bedford. In a few days, the Prime Minister of England will grace it with his presence. Now you must help me. Do tell, how should we decorate the house for the Prime Minister's visit?'

Smith sounds as doting fathers do, coaxing their daughters back to mirth with apple caramel pudding.

Mrs Kitten sighs. 'You have a mortgage on the Trimbridge house. You have a mortgage on the Tucking Mill house. There is the monthly bill to tackle at the Swan Inn. The annual rent and charges for this Buckingham Street property. What else should I list out? Those who you owe send letters each month without fail.'

Smith changes the topic. 'I have received some great news, Mrs Kitten. The Society of Arts has posted an advertisement. The one who can draw and submit a map of the complete mines of even one of the three regions of England, Wales or Scotland, will be rewarded fifty guineas.'

'And do you think you will be able to give it to them?'

'Who else but me? Do you know what the Duke of Bedford says? To whoever he meets? "Smith is God-gifted. He can take a look at soil anywhere, anywhere at all, and tell what is there under it."'

'That logic does not sound too complementary, Mr Smith. If word of this goes around, crooks and mine mafia groups could make you their target. Perhaps, they already have done so… And you are completely oblivious to who is good and who is not.'

Smith smiles innocently, 'Guilty as charged. But wouldn't you agree, no one, almost no one, knows rocks the way I do?'

'Well, if such happiness is enough for you, then who am I to say anything? Does your wife have nothing to say about all this? I haven't met her ever. What is she like, this girl, who tolerates you as she does?'

Mrs Kitten realises that she's said too much. She hadn't really thought it through. But she notices now how quickly Smith's expression has changed. He looks like a stricken man. He suddenly seems to be concentrating completely on his ink pot, and the pen in it. This man falls silent, always, if his wife is ever mentioned. He turns to stone, petrified, needing time to come to terms with his present situation again.

Even today it takes him time to return to normal. He smiles wanly, and says, 'Let the poor thing be, why drag her into this? I must be on my way now.'

'There is something else. In truth, I have felt so unhappy since I heard it, that I just went on babbling senselessly all this while.'

'What can be worse now? Well, out with it then. I feel delighted with bad news. The Almighty has chosen me, William Smith, especially for all sorts of bad news. God's choice is incomparable. Yes, Mrs Kitten, let me hear you out as I enjoy my coffee.'

'Your Trimbridge house has burnt down to the ground.'

'Wonderful! And is the caretaker alive? In case not, then has his family demanded one thousand pounds as compensation?'

'The caretaker is unharmed. He has somehow managed to save your documents. But not the fossils. He wasn't able to shift all those large heavy boxes on his own, and the fire spread very swiftly.'

'All the fossils? Gone? My fourteen years of toil. How long did they burn?'

Mrs Kitten holds herself stoically.

'A wooden house, it barely took one and a half hours. It was struck down by lightning.'

'How forceful it must have been, the power of the bolt! To reduce to ash Williams Smith's fourteen years to nothing. If only men could possess an iota of this power.'

'You will visit Trimbridge, won't you? Should I arrange for a carriage?'

'What's the point now? Either I'll bend my head down at the ashes or cry over the carnage. I do not like either.'

'It's not only about the mortgage. You must claim the insurance as well.'

'But there is no insurance.'

'What! Why?'

'I hadn't thought it would come to this. After all, the house was surrounded by rocky hills and a river. Since there were no trees around, where was the possibility of a fire? I hadn't thought of lightning though.'

'There is nothing else left for me to say! In any case, what can I even say to you?' Mrs Kitten gets up to leave.

'Don't leave yet, please. Do read me the letter. There is nothing personal that I need to keep from you.'

Mrs Kitten opens the letter.

'Who is it from?'

'It says, "your beloved Liz." Who is this Liz?'

Smith's face glows bright. 'My younger sister, Elizabeth. Has she written it herself? Please, do read it.'

My beloved brother Willy,

You will be surprised surely to receive this letter. How could your little Liz write a letter like this? But it is not me. I wish to say so much, and yet I cannot write. My health is so poor that I have no strength to write even two straight lines. You just left and went away from our uncle's. And we three continued to grow. We did not grow as big as you have, and yet I grew enough to have the others think of getting me to do all our house work, and to get married. I was married off to a Mr Phillips from a Welsh family. I am lucky to be married to a wonderful man who loves me immensely. We are now blessed with three children. My girl is the sweetest in this world. She is four. She looks so well after her little brother. My youngest, a boy, is just two months old.

And my oldest is John, a seven-year-old now. If only you were to see him, it's beyond all my belief! It is you, a replica, that stands before me. It is you and more handsome. And he is just as you were, spends all his time, wandering around the jungle, staring at rivers, looking at stones. He collects rocks, keeps making notes. He has learnt to read and write all on his own. Do you know that there is now a school in our village? John studies there. The schoolmaster hopes for a lot from him. He thinks John will grow up to be a wise man. And then there is my sweet, sweet Annie. The youngest is Jenkins. Both John and Annie love to read. I do not know how far I can help Annie study, but I have hopes for John. I have told him that if he can study more than my brother Willy, then I will know that he has mettle.

But I do not know if I will be alive to see that day. Even Annie's father does not keep too well. I am scared, Willy. I so wish to look upon you, my brother. You are still indebted to me for my seven shillings. You must come now and give it to me yourself. I can speak no more. I am beginning to cough again. Someone in the village told me that you have bought a huge mansion next to the Thames. There's a picture of it in the newspaper. I was so lucky it had your address. I could send this letter. The others are piled up here, addressed but unsent. I'll hand you all your letters when you come. Tiny letters that my fever never permitted to become longer. The fever comes often now. Come once please, my dear dearest Willy. Promise to God, you'll come as soon as you receive this.

Your beloved Liz.

'What have you decided, Mr Smith. Is it Cambridge or Churchill?'

'Not Churchill. Liz's husband lives in Marden. It's in the Oxfordshire county.'

'Let me know where you wish to go, and I will arrange your bags accordingly.'

'No point visiting Trimbridge now, Mrs Kitten. It will be better to go to Marden.'

'Yes, that sounds better. I will pack your clothes. You must carry a gift for your sister. And a few toys for the children.'

'Yes, yes, absolutely. But what should I do about Cary?'

'Think of him another time.'

'But I did promise to show him the maps. I say, please pack my bags and call for a carriage. Here are ten pounds. Do pick up gifts as you think fit. I'll leave by this evening.'

'Don't be too late. You must leave while it's still light.'

'I will do so without fail. I'll be back soon.'

***

That evening Mrs Kitten waits till 7. There is no sign of Smith at all. She finally sends a man running to Cary's office. Smith has left Cary's at 2 that afternoon. But no one knows where he has gone.

Mrs Kitten does not show any further eagerness to look for her employer.

YEAR 1820, LONDON

In 1807, almost thirteen years earlier, on 13th November, a dining club was inaugurated in London's Freemasons' Tavern. Humphry Davy received an invite to the programme, which mentioned that a few notables were starting a small Talking Geological Dinner club, and he was graciously invited to accept a position as one of its members. On its inaugural, the Club was named the Geological Society of London. Of its initial thirteen members, only one, James Lewis, could perhaps be called a geologist, certainly none of the others. And Lewis only on account of being involved in a business that dealt with minerals. Of the rest of the members four were doctors, three chemists, two publishers, a Mr William Allen, the partner of a pharmaceutical company Allen & Hanburys, and a minister of the Unitarian Church named Arthur Aikin who was gaining popularity as a church administrator, proficient in raising continuous questions when citizens became involved in matters of the church. In straight terms, apart from Sir Humphry, not one member was truly a notable who could be said to have achieved any recognisable degree of fame or to be an established name in the world of science.

One is also hard put to find connections with anything remotely scientific with the person who was elected President of the Society. His name is George Bellas Greenough. A natty, polished youth of thirty, with not a trace of anything rural or inelegant. If put on a dais ever without notice, he can speak extempore on any given topic for at least five minutes. The audience is inevitably mesmerised, be it with his fashionable attire, or the manner in which he turns his neck, the flash of his diamond ring or even by his blue eyes. Greenough is a rich man. In fact, he is now more in the news since he built a huge mansion in London's Royal Park Regents. If he gives a miss to a party thrown by a duke or an earl, there is bound to be gossip. And even if there are doubts about his lineage, no one dares speak openly about it. Cynics, of course, maintain that Greenough's grandfather, that is his mother's father, was a quack, selling a digestive under a fake label, and who made pots of money, and therefore, a name. It is even believed that Greenough is not an actual family name. According to the rich grandfather's will, if his beloved grandson wished to inherit all his wealth, and his entire property, he'd have to rechristen himself Greenough.

A very disturbing matter had reared its head even while the first meet of the Society was on. Joseph Banks, President of the Royal Society, objected to the independent nature of this new Society. According to him, it could not be allowed and the new chapter would have to merge with the Royal Society. Only Humphry Davy agreed with what Banks was proposing. That day Banks withdrew his membership. A while later, once a few sessions had been completed, geologist James Hall, a Neptunist, joined the Society. By now the scientific community was divided into two groups, broken up by their differences over the source of the earth's rocks. The Neptunists believed that the rocks had formed in the belly of our oceans with the hardening of silt and clay sediments across their edges, layer after layer. On the opposite side, the Plutonists were convinced that the rocks had formed due to the cooling down of hot molten magma on the sea bed which had happened as a result of volcanic eruptions. There was no end to these debates. No one knew who would have the last word.

Today, it is the 12th of April, 1820. The bell on the main door of the Buckingham Street house begins to clang. Mrs Kitten feels strangely worked up at its sound. Even a few months earlier, she would not have been so anxious about it ringing. Who knows what ails her nowadays? The sound of the bell sets her heart aflutter. She opens the door. There are four men at the door. Two of them she recognises, they have visited this house on a previous occasion. She recalls that day and feels a sharp pain rising in her heart. But she asks, her voice calm, 'Good afternoon, are you looking for someone?'

The most handsome of the lot steps up and curtsies. When this same man had come previously, he hadn't bothered to greet her at all.

'Good morning, Mrs Kitten. I am Greenough, the President of the Geological Society. Is Mr Smith at home?'

Mrs Kitten smiles archly and replies, 'No, he is not.'

The men do not react to this in any way. It's obvious they know that Smith is not at home now.

'That's a pity, although it doesn't really matter. We don't need him for the work we have come to do.'

Mrs Kitten continues to stand her ground. Only her eyebrows seem to register this comment.

The second man, he too familiar, moves up and says, 'I am James Hall. The last time I had visited, I promised to come back again. Here we are now.'

Mrs Kitten remains unmoving. Again, only her brows shoot up.

A third man now walks up. Mrs Kitten doesn't know this one. He looks somewhat like a guard or security person.

He says, 'What are you waiting for, Mrs Kitten? These are respectable citizens. They never visit anyone uninvited, and they have graced this house of their own volition.'

Mrs Kitten looks at the maple tree standing tall at a distance. 'I am acquainted with them. They visited about a year ago.'

'Then what's stopping you now? Just let them go in.'

'I still don't know why they are here today.'

'On work, of course,' says the beefy man, trying to act polite. 'They have come again to take a look at your employer Mr Smith's collection.'

'My employer is not here just now. You may come back when he returns.'

'You are making a mistake, Mrs Kitten. As we mentioned right at the start, Mr Smith's presence is not required. We will simply take a look at his collection and leave.'

'I am sorry. I cannot show you anything without his permission. We will meet another time.' Mrs Kitten is about to shut the door when the beefy man wedges his foot in. He smiles slyly.

'Ha! What's the rush, Mrs Kitten? Wait a while. It's such a pleasure to meet a loyal employee such as you. I can't help but praise you.'

'Thank you. Goodbye.'

'As I just said, there is no rush. We bear a letter of authority from the Parliament. We have every right to look at Smith's collection. Please read this letter. You are clearly educated.'

Mrs Kitten reaches for the letter. It says:

The entire collection of Mr William Smith that is at present housed in his Buckingham Street residence, is to be forthwith taken entirely under the care of the British government. To check out this collection, a delegation of four personages will visit William Smith's Buckingham Street residence on the 12th of April of the year 1820.

Mrs Kitten does her best with one last attempt. 'But there is a board hanging outside. It says intruders will be prosecuted, that is if you paid attention to it.'

'You are mistaken, Mrs Kitten. That board is authorised by us. So that a special gent does not come in here. You surely know who we mean.'

Greenough begins to mutter to his mates, 'Why do we even need to engage with this woman? Let's get two policemen. It's all we need.'

The beefy man intrudes, 'Mrs Kitten, everything in this house belongs to the Society. It is their property. We could call the police, but we do not behave badly with women. And the best thing is that you are receiving your monthly pounds, aren't you?'

'What are you saying?'

'There is no need to act dumb. No one knows where your employer is, but you are still receiving money. Where do you think it is coming from? Now enough is enough, step aside and let us in.'

Mrs Kitten feels paralysed all of a sudden. She has turned to stone, her hands and feet are not moving, her ears feel blocked. The four men push her aside and barge in. The entire stairway resounds with the pounding of boots, the same grating sound she'd heard when they had visited last.

Mrs Kitten quietly feels the keys in her pocket. A voice floats down, 'Hope you're fine, Mrs Kitten. Please come up and unlock the door to the collection room.'

Mrs Kitten's hands are shaking horribly. She's finding it difficult to move.

The beefy man yanks the keys away, 'Why are you stalling us? These are important men, each minute of theirs is valuable.' His voice is now changing, it is almost threatening. 'Switch on the lights fast now, don't be tardy.'

What a change in just a few moments! Now it's a masterful tone, it's demanding. In all these years, Smith has never ever spoken to her like this. It is as if these men are quick-shift artists, who talk to different people in different ways. Mrs Kitten finds the candles and lights the lamps one by one. The damp of the room hits them now, the windows have been left shut for a while.

The men immediately open the showcase and begin their search. Mrs Kitten has recovered a little now and is scrutinising the intruders carefully. They aren't here for the fossils. They are ignoring the labelled artefacts, not even bothering to read their contents or pick them up. They are hunting for something else.

'Hey, Farey, hadn't you mentioned that we would find them here?' The handsome fellow speaks up.

Hmm! So the burly one is Farey. A fraudulent type, more like a broker. She remembers Smith, weeping, his face wet with tears, his voice, choking. 'The same Farey, a man I considered my dear friend, whom I taught so much, showed everything to, has betrayed me now, Mrs Kitten. He smuggled away all my maps to Greenough. I gave my notebooks to him to read. I had no idea when he made copies of those. I had never imagined that this would be the price to pay for trusting someone.'

Mrs Kitten looks daggers at Farey. She's standing close to the door, where it's very dark. The men cannot see her expressions.

Greenough says, 'Ask that woman. She might of course not know anything. She's an ordinary clerk, after all. But try at least.'

Farey approaches her. 'You seem such a fine worker, Mrs Kitten. We need someone like you for our work, efficient and smart. At a salary of three and a half pounds.'

Greenough mutters again, 'Just get to the point, will you? You can't always be trying your tricks.'

He has spoken in a whisper. But Mrs Kitten knows his words. She lives with her son, her only son, who cannot speak. Mrs Kitten knows well how to communicate by reading lips, having learnt this skill from her son's teacher, Mrs Stanley.

Farey speaks again, 'How many maps were there, Mrs Kitten? Where are they now?'

'Maps? What are those, sir?' Mrs Kitten asks.

'Maps, as in large sheets of paper with drawings and scribbles, colours filled in on big sheets.'

'Oh, there are many of those here.'

'Really? Where? Do show.'

'In the living room. Beautifully arranged. Not my master's though, they belong to the house owner, framed in wood.'

'The owner of this house? What kind of pictures?'

'Beautiful ones. There's one of a girl standing, holding a kitten in arms. Her gown is all white, a pink flush suffusing her entire being. And then, there is the one with two beauties standing by a waterfall, calling out to all the birds flying above. Their arms, so long and graceful as if they themselves are birds, about to fly off freely.'

'Pfft! Ignoramus! Labourers, subsisting on daily wages. Tell her to stop.' Greenough's beautiful face is scrunched up with irritation.

Mrs Kitten is quiet again.

'Where can those maps be? Smith was not allowed in here after his release from prison. Or did he sneak in, eh, Mrs Kitten?' says Farey.

'Well, if he did, I didn't meet him. But who knows, in the dark of the night, like a thief…'

'Have you seen any papers around here, important ones perhaps?'

'There was a notebook, with notes on his fossils. Mr Smith carried it with him, but that was almost a month before his imprisonment.'

'Well, that's strange. The maps were here I am sure. Smith told me so several times,' Farey is not able to fathom what could have happened.

Mrs Kitten speaks up again, 'Respected Mr Greenough, when you had first visited, had you seen these maps or whatever name you call it? That day you had spent a good amount of time, looking through the fossils, hadn't you?'

Greenough falters now, 'No, I mean, yes, yes, indeed. But we hadn't seen any maps that day.'

Mrs Kitten is dying to laugh. This same man had spent a total of five minutes at most in this room while men like Sir Joseph Banks had spent an entire week, just looking around at the collection.

'Fine then, Mrs Kitten, please send word if you come upon the maps,' Greenough says, slightly dejected.

'Of course, I shall. But I am not educated enough to understand what you are looking for. If you send me a sample, it would do nicely.'

'That's no problem. Farey, send her one of mine.'

The men begin to file out. Mrs Kitten speaks again, her voice soft and gentle now, 'When will you send one of Mr Greenough's maps, Mr Farey?'

Farey walks back and plants himself in front of her. He is almost breathing down her face.

'Very soon, Mrs Kitten. Do not worry.'

He casts a long burning look at her, and then goes down the stairs, one heavy step at a time.

After she has bolted the main door to the house, Mrs Kitten trundles up to the fifth floor office once again. She takes out a sheaf of papers from a drawer. Are these safe here anymore? These thieves could be back again, thieves who pretend to be wise and giving, but who attack in broad daylight. In truth, they are worshippers of the devil, with the power to crucify as the devil did to the son of God. They can sow the seeds of conflict between brothers like he did to the sons of Abraham. It is no problem for them to seize all of Smith's maps.

She can never forget that one day even if she tries as hard as she does. Smith had come running into the house, bounding up the stairs his face wet with joy. He had gasped, 'Tomorrow some notable people of the Geological Society will come to visit. What should I wear, Mrs Kitten?'

Mrs Kitten had dug out a shirt and a coat, but found no ties for him. The next morning, very early, she came back carrying a tie borrowed from her husband.

Some men had arrived around 10 in the morning. Mrs Kitten had opened the door and presented the two visitors with bouquets. Smith had whispered to her, 'That pretty boy is the President of the Geological Society.' The other man, who Smith knew very well, was James Hall. They had come to take a look at his collection. 'Just look at how keenly they are looking over everything. James, of course, knows quite a bit about my work.'

Smith had then bounded up the stairs, his excitement palpable. He was talking too much, much more than he ever did, and much more than was required. Yes, he was delighted by their presence. It was Smith first on the stairs, followed by Hall, and then at the end, the princely man, Greenough. Smith was blabbering away, nineteen to a dozen, and Greenough was looking around, disgusted, a handkerchief clamped to his nose, his hands moving swiftly away from the railings whenever he was about to touch it. He clearly wished to be far from what he thought a grimy, dirty place.

Inside the collection room, Smith busied himself with opening one shelf after the other. He began by explaining how to calculate the age and geological set-up of a rock from the fossil it contained. A sedimentary rock with fossil embedded in it is itself a marker of a specific geological age and reveals the geography of the time when it was formed. So mesmerised was he, all over again, by the wonder of these rocks, their age and uniqueness, from across the expanse of England, that he seemed to be conjuring up the formation of each rock layer of the geological past. The pretty man was pretty unconcerned, far more involved in staring up at the ceiling, as if to detail each layer of dust and muck hanging downwards. Smith opened a map in front, pointing one layer after another to them, while he, the pretty one, was telling Hall, 'Can you see how rustic he still looks? One cannot see even a trace of London-polish in your Mr Smith.' And as Smith pointed to the spots in the fossils that indicated their age, the man had continued, 'Who knows where he dug out that shirt from. And his stale smell, God knows when last he took a wash. Even my pets are cleaner than he is.'

James had found this funny and guffawed.

Smith had fallen silent and stared helplessly, 'What is it, Mr Hall?'

'Nothing, nothing to be concerned about. We have understood whatever you're trying to tell us.'

Smith continued naively, 'I've only told you about two layers so far. Wait for the rest, they are extremely interesting.'

The beauteous being had spoken up, 'We are done, Mr Smith.'

Smith said piteously, 'How is that possible? I've hardly even started.'

Greenough had cut him short, 'We don't need to see anything anymore. I just told you, we have seen enough.'

Smith had stared at James Hall, sadly, like a man drowning, clutching at hope. 'Did you not like this collection, Mr Hall? Should I go over the details again?'

Greenough checked his pocket watch, 'I don't have more time, Mr Hall. I need to check on the arrangements for this evening's party.'

Perhaps Hall was a little embarrassed, feeling a tad bad for Smith. He had offered, 'You see, we have a Society meeting this evening.'

'Meeting-cum-dinner,' Greenough added. 'Members need to pay two shillings each for the dinner. Come now, Mr Hall. We'll be late,' and walked off even as he spoke.

James Hall had placed an arm across Smith's now-drooping shoulder, 'We will come back, Smith. Bye for now.'

Smith had continued to stare at their retreating backs, frozen. Mrs Kitten had brought him a hot cup of coffee. Smith had asked her then, 'Why did they go away, Mrs Kitten? Do I not sound polished enough, like a gentleman?'

Mrs Kitten still feels rattled when she recalls that day. Now she clutches the brown packet close to her chest. Would it be best to reach these maps to John Cary, to his office? Four years back, a few of Smith's maps had been published from Cary's publishing house. Many more maps are waiting to be printed. Smith has made each one of these maps of the entire sweep of England, Wales and Scotland by travelling across each inch, digging into its layers. Maps of different scales, which this mad man has drawn, wandering alone, endlessly with no hope of reward. And no one knows, no one will ever know! How is that possible? How can this be allowed? Only Cary will understand, only he will know what to do. But, then, how can she be sure that he has not sold himself out. There seems to be not one soul who can be trusted fully. Who should she take the maps to for safekeeping? These maps are the sign of the pure devotion of one man alone, his tireless toil for sixteen long years!

Oh, Dear God! If such great power resides in the devil himself, how will man be helped by calling upon your name?

YEAR 1808, OXFORDSHIRE

In most villages, it is usually not too difficult to locate someone's house because almost everyone knows everyone else having lived in the same place over generations. The first man walking by with a sickle in his hand replies when asked, 'There are two Phillips in the village. One, a farmer and the other, employed in an office.'

Smith does not know Liz's husband's profession. He thinks for a few moments and says, 'I think, yes, he works somewhere.'

Now the man says, 'Walk straight from here for two miles. You'll come upon a forested patch. Halfway through this area, there's a mud path to the west. Keep walking along this till you arrive at a village. Worker Phillips lives there.'

Smith is now walking along the mud path to the west. The walk is unending. Where is the village? Smith doesn't mind walking, it doesn't tire him out. But he hopes against hope that he hasn't made a mistake. Maybe he should have first enquired about Farmer Phillips. The jungle patch is full of dry trees. Smith has come upon a narrow streamlet now. He has his fill of water. But there's not a person in sight anymore. No, no, he had spotted a man along the way some time back, staring up at a beehive so transfixed as if he had seen such an object for the very first time in his life. The man cannot be seen anymore, but the buzz of the bees is still loud and clear. Smith turns around and begins walking back the way he had come. It doesn't matter how much time it takes, but he must go in the correct direction. He can't wander around confused anymore. It will soon be dark.

He sees the man again after walking a little more than a mile. The man's still staring at the bees. Why does it feel as if the man is not an uneducated villager? He's clearly not waiting to collect honey. He is studying them, the bees swarming around the hive. He has changed his position from where he was standing earlier. He has chosen his spot carefully, around ten to fifteen steps away from the bees, where the bees are not disturbed and are not troubling him either. The man does not seem overly concerned about the large dangerous-looking swarm.

'Hello, Mister, please can you tell me something?'

When he sees Smith advancing, the man puts a finger to his lip and gestures that Smith move back. Smith halts and then steps back slowly. The man moves back too and gestures again that Smith move back further. And further back.

Shuffling back like this, perhaps fifty steps or so, they come and stand at a particular point.

'What is it, tell me now.'

The man has a heavy voice. It has a touch of the village, and yet a peculiar polish as well. Almost as if the man recites poetry often.

'Where does Phillips live in this village?'

'Are you the same man who walked ahead a while back?'

'Yes, of course. Someone directed me past the jungle patch. The village is Marden. Apparently, Phillips the worker, lives there.'

'Indeed, the man used to work in the accounts department.'

'Used to? As in he doesn't anymore?'

'Nah! It's been a while now.'

'Does that mean I've come to the wrong place? It seemed from Liz's letter that her husband was doing well.'

'Liz, as in Elizabeth, right?'

'Yes, Elizabeth, yes. Do you know the family? They have three children, the eldest is about seven, or so. A fine lad, very intelligent I believe. I'm forgetting his name. It was mentioned in the letter.'

'It is John. Walk with me, I am headed that way.'

The men begin walking again. Smith hardly talks. This other man is more silent than Smith. His answer to everything is 'hmm', or even shorter. But it is clear now that he is a teacher in a school somewhere close by. Perhaps, as is common in villages such as these, men like him think no end of themselves, surrounded as they are by uneducated bumpkins.

When Smith offers the information that he is Elizabeth's older brother, the man says 'hmm' again, as if he is already privy to that information. It is quite annoying to see such know-it-all behaviour in men. But Smith has no option other than to walk along with him. It is important to cross this forested wilderness quickly, and more urgent to get to Liz's home in the middle of this godforsaken place. At least the man has some education. At least some conversation is possible.

After they have walked around another one-and-a-half miles, the man brings Smith to a large mound of stones and points out, 'It's not far from here. Right there in front.'

'What do you mean in front? I can see no houses there at all.'

'Can you not see a few people gathered there?'

Smith wipes his sweat-reddened face on his grimy shirt and says, 'I thought you were an educated man. And yet you haven't understood what I meant. I wish to get to Phillips's house. You have brought me to this dry deserted place. I can see no habitation around at all.'

'You are right, there is no habitation around. People come here only when they have no other way out. Although the front part there is not so dry. It is a graveyard.'

Smith is livid. 'You are a strange man, Mister. What do you mean by bringing me here? How do you profit by troubling a tired and hungry traveller like this?'

Suddenly, Smith thinks back to the last word the man has uttered, 'Why a graveyard?'

'Because this is where they are holding the 40th day memorial service for Elizabeth, the wife of Phillips of the Welsh family. I thought it best to bring you here first.'

Smith slumps down against a large rock. He is now rubbing his right palm against it. He can't stop. He seems lost. The man stands by silently. He does not interrupt. Men mourn in different ways, to hurt oneself is also to mourn.

The man waits for quite some time. Then he says, 'We shouldn't delay any further, Mr Smith. The prayers will begin soon. Come.'

He extends his hand. Smith looks up at him, his eyes blank. 'I made a grave mistake. I didn't think she was so far gone, my little sister. If only…in truth…'

The man pre-empts him, 'You were too busy, were you not, Mr Smith? The letter had been written eight months ago. I wrote it for her. Even then she couldn't speak for long. She was coughing blood. She was alive after that for at least another six months. If only you had found the time. But there will be enough time for regrets. Come, you must first be a part of the prayer meet. Get up, Mr Smith.'

It is afternoon by the time the prayer meeting concludes. Everyone sits together under the tin shed of an ordinary village church. The shed is a relief. Only a few people have collected. It is just two small groups, a total of fifteen people. People are stealing glances at Smith, every once in a while, even as they keep talking to themselves. He is clearly the topic of conversation at the moment.

Sitting alone, Smith has by now downed two glasses of juice. He is not feeling as exhausted anymore.

Smith wonders if he should leave now. He would have liked to meet Liz's husband. Which John is it in this crowd of several Johns? No one has introduced the two. Perhaps, John does not wish to make his acquaintance anymore. Perhaps, it is all for the best. He doesn't deserve to meet his family, or get to know them. Where are his other brothers? He can't recognise them in the crowd. Have Dan and John not received news of Liz's death? Their sad, broken little family. How can death be powerful enough to glue those broken bits back together again? He could never know what it meant to truly be a family. It's an apt name that he has been given, Strata Smith, sandwiched between layers of rock, a life full of rocks, that is all. Except his memories, he carries no one along. But God has been gracious, at least he has his memories. Why did he even come here this far? One more tragic memory added to his already long list.

Smith shuffles out of the church.

'Are you leaving, Mr Smith?'

The same man, the village school master. Smith hasn't seen him since coming to the church. He may have been busy looking after the guests.

'Yes, Mr…'

'Winson. Bob Winson. You can call me Bob.'

'Thank you, Mr Winson. Please forgive me for being rude a while back.'

'It is not important. But you cannot leave right now. There are five guests from Churchill. Why don't you go with them? After her husband's death, Mrs Phillips began living in Churchill.'

Smith sinks down on a bench close by. He looks infirm, like an unwell man.

Winson sits by him. He's saying something softly, almost whispering, 'When I wrote the letter for Mrs Phillips, her husband was still alive. But he wasn't at all in good health.'

'Why didn't you write to me?' Smith barely whispers back.

'She did not wish to send you bad news, she had hoped that you would visit, and see everything for yourself. But later, sometime after Mr Phillips's death, a letter had been sent to you. But there was no answer to it.'

'I received no such letter. I was not in London these last eight or nine months. I have been travelling. I've come directly from my travels.'

'But you had received the letter written eight months back, had you not, Mr Smith?'

'Hmm,' Smith stares at the ground, afraid to meet Winson's eyes. How can he excuse his own blunder? That other day, he had promised Mrs Kitten that he would leave for Liz's as soon as he got back from Cary's office. But he had met someone there, at Cary's, a man who was waiting because he had known Smith would be visiting. Mr Morgan, a mine owner of the Combe Hay area. The layers of coal in his mine had sunk in very deep and then had suddenly bent like a hair pin. Mr Morgan just could not fathom how much the layers were bent or how deep one needed to dig for the next layer. He wished for Smith to carry out the survey. He would pay him handsomely. But God is his witness, Smith had not agreed to this assignment for the money. He had had to halt just such a job in Somerset some time back to begin surveying the canals. But his heart had always craved to try out this challenge. On that day, while agreeing to Morgan's proposal, Smith had no thought of his little sister. He had not even demurred once when the mine owner asked him to go along in his carriage. Not even asked for a few days of leave. That too was greed, an obsession, a sharp desire, just like the need for liquor or gambling. Just that his need cannot be explained or even understood.

Smith asks again, his voice faint, 'Who lives in Churchill?'

'Your nephew and niece, Jenkins and Annie, are with your two brothers, John and Dan now.'

'I see.'

'Come then, I'll acquaint you with the group from Churchill.'

'I won't go to Churchill.'

'Fine, then, come with me to Marden. Your eldest nephew is still there, with his uncle. Just wait a while please, Mr Smith. We'll start for Marden in a bit. Please, please do not run away. And in any case, there is no way back to London tonight. You will have to stay in the village.'

***

The house is falling apart. Termites have almost eaten away the windowsill. The foxgloves trailing across the fence and the daisies in the ground are all dry and droopy from neglect. Does this portend something terrible again?

Winson walks into a room and comes out in a little while.

'Mr Phillips's younger brother is extremely unwell. He is suffering from flu. He is unable to breathe, but he wishes to speak with you. Would you please come in?'

Smith goes in. He himself is thirty-nine years old. Liz was five years younger. But this face that gapes back at him, pasty with illness, and its blank cloudy eyes, looks far more gaunt, much older than seventy. If this is Liz's brother-in-law, how old would her husband have been?

But even so Smith nods, with just a hint of a smile. Then he looks away. The man tries his utmost to smile back. But it is painful to even imagine that it is a smile, the skin on his face, wrinkled and stretched over protruding bones. The man is too weak to speak. The room reeks of death. He is not going to live more than a few days. The man reaches under his pillow and pulls out a tiny wooden box. He passes it to Smith. Smith knows well what the box contains. He does not need to open it. He wordlessly clutches it to his heart, and flops down on a chair, stunned by the turn of events.

Now the man holds up a piece of paper. A hand-drawn portrait, in all probability a portrait of Liz. Smith has not seen Liz as an adolescent, let alone as a young woman. He wants to look at the picture more closely. But the man refuses to hand it to him. He folds it again and slips it back under his pillow. But he does grab Smith's hands in his, presses it with all his might. But the effort makes him wheeze again. He's grunting now, tears streaming down his face.

Smith doesn't know what to do. He can't breathe either. He wants to snatch his hands away and begin running. He wants to run and run and not stop until he has left this village, this forest, this situation far behind. He will stop only when he comes to familiar haunts.

But time seems to stand still. Very soon, however, Smith finds relief. The man has slumped over. His grip loosens and his skeleton arms now rest on the bed. Smith is free, free again.

Winson speaks up, 'Come with me, sir. I live close by. Please stay the night there.'

After dinner, Smith unlocks the tiny wooden box. It's crammed with bits of paper, with only one or two lines written on them. One can hardly call them letters. There are no envelopes. After all, Liz had no address to send them to. It's a childish hand, crooked, words sitting on top of each other.

How are you, my brother? I am well.

Another says: *Do you remember that day? I remember it too well.* But nothing about which day, or what had taken place then.

I've had a son. He's such an angel. Come, come please.

Bits of memory, just memories.

There is a longer one, three lines long.

Do you not think of me? I know you have suffered much. But hasn't the pain gone now that you are such a big man? I am so proud of you, my brother.

'May I come in, Mr Smith?' It is Winson.

Smith opens the door. Winson walks in, 'Your brothers had visited earlier. I had requested help for Mr Phillips's brother. They gave what they could.'

'How much would you need, Mr Winson?'

'There is no need, Mr Smith, he will not live long now.'

'But we must try, mustn't we?'

'He doesn't wish to live. He doesn't take his medicine anymore.'

Smith does not know what to say in response.

Winson speaks up again, 'Do you know what he was trying to tell you when he held your hand?'

'No.'

'Your sister's eldest, a very bright lad, lives here. The daughter Annie is also a bright spark. Your brothers have taken her and Jenkins with them. But what will become of this one here? I have taught him all I could, and the villagers have donated what they could. But it may not be possible much longer. Your sister dreamt for John to study in Oxford.'

Smith chuckles, 'Have you ever heard of a farmer or a smith's son studying at Oxford? I'm not known for my education either. I wander around, rolling from one place to another, working when work comes my way. I have a wife at home, ill and needing attention, which I cannot give. I have no time for myself. Where will I take this little boy, how will I look after him? I'd rather leave some money for him with you, for his food and boarding. Let's think of his education later.'

'Well, in that case, let God take care of us. Good night, Mr Smith.'

Smith stands up, too. 'Good night, I'll be on my way early tomorrow morning. I'll visit Liz's grave before I leave. Please do not bother with arranging for breakfast.'

'As you wish. Please, do rest up.'

***

The next morning, Smith walks to Liz's grave. It is silent and deserted all around. There is a cool breeze. The grave is not yet dry. Someone's planted roses around it. But Liz loved lilies, not roses. Smith places seven shillings next to the roses. There is no one around at the moment. He can cry his heart out. Smith looks up, just to make sure. And spots a little boy, sitting atop another grave. The boy is wearing only a thin shirt against the bone-wrenching chill of the morning. His shoes are battered, his toes peep out. The little one is staring at Smith, his eyes wet.

Smith gestures. The boy walks up to him. In his hand is a stone. Smith says, 'Poundstone?'

'This is not a stone, it's a marine fossil. Clypeus. The villagers know no better. That's why they call it a stone.'

'You seem to know a lot. Did you read it in a book?'

'Yes, in a book. You're the one who wrote it.'

'I wrote it? You know that, too? Do you know who I am?'

'Uncle Smith?'

'Are you Liz's eldest?'

'Yes, I am John.'

'Why aren't you wearing something warmer in this cold? You may catch a chill and a fever.'

'I don't feel the cold.'

'But look at the dirt in your hair. Do you not take a bath?'

'I do, when I feel like it. If we sit in dirt, we are bound to get dirty, right?'

A child, and yet how well he explains himself. Delightful. So very mature. Grief makes mature beings of us all. Was Smith himself like this as a boy?

Smith asks gently, 'Why do you play with dirt?'

'I like to look for stones.'

'Do you recognise stones? Tell me about that one there, what stone is it?' Smith points to one among a mound.

'That's chalk. It will emit carbon dioxide if acid is poured on it.'

Smith had not expected such a smart reply. He is stumped.

He questions the boy again, 'Have you seen this happening?'

'Yes, I have also seen the geological map in which you marked this one as chalk.'

'Where did you find my map?'

'Mr Winson showed it to me.'

'Can you read maps?'

The boy gives a crooked smile, as if to say, are you stupid or what, to ask such a question?

But he says, 'Mother used to say, you are a man with great knowledge. But there is something you have not yet managed to find, which I think I might.'

Smith's eyes pop out. He wants to laugh, but he doesn't. The boy is speaking very seriously.

'Tell me about it.'

'You keep saying this stone is older than that, and that has come even before that other. But you do not give us the actual dates or ages. How does that help? We need ages and dates for sure.'

Now Smith bursts out laughing. The boy looks at him annoyed, his brows wrinkled. He does not like Smith's behaviour.

Smith holds himself back, 'How will you do it? Let's hear that. Have you already built an instrument?'

'Not yet, but I will. I need to study first.'

'Of course, that's true. It is very important. Come on now, let's get moving. Let's go home.'

John walks ahead. Smith follows and comes to stand by him.

'You are quite a handful, aren't you? How full of dirt you are, your hair all dry.'

Smith pretends to dust the boy's hair. But he takes in the scent. A young head, full of sun-baked hair. It's a familiar scent. Perhaps, he is taking in the scent of his own self.

The boy asks, 'Why are you looking back so often, Uncle Smith?'

'No, nothing. It's just that I had forgotten to tell your mother something. I just whispered it to her.'

The boy stands akimbo. 'She's under the ground now, sleeping. How can you talk to her?'

Smith smiles, 'You don't need to know everything all at once. Let's go, see what can be done with you.'

'Done?'

'Bah humbug! So many questions. Your mother said you were like me. I can't remember if I chattered so much in my childhood. I will now have to take this chatterbox with me. And then, he has to study at Oxford. Do you know how much you need to study? You will have to listen very carefully to me.'

Liz's boy is indeed very delightfully restless. He is skipping ahead now, almost ten steps ahead of his Uncle Smith. His shoes are throwing up clouds of dust behind him.

The little boy does not care.

YEAR 1822, LONDON

A week back Fitton had taken an appointment to meet the person he has come to see today. This person has asked for a meeting in his office room at the British Museum. It's morning right now, the clock shows it to be 10:20. Fitton will be asked to step in ten minutes later. He collects himself. He has been becoming more emotional about this matter as each day passes. But it does not suit a man of his social standing to show undue emotions in his dealings. Particularly today. He must be restrained and extremely calculated in what he says or expresses to the man he is about to meet. But what he observed the day before, at the British Museum, while on a personal visit, has not allowed Fitton even a wink of sleep last night.

Officially, the inroads into scientific study began some two hundred years ago. But the waves of scientific discoveries began gathering force only some fifty to sixty years back. And with it began the play of dirty politics around scientific advancements, as also the need to be recognised by the world, to receive the respect and honour of everyone around, and to participate in its garish celebrations. This intoxicating behaviour spread across slowly, but surely, and so did the desire to plagiarise and steal, along with the mentality to deprive others of their deserved rewards to further one's personal interests. These are wily people, cruel and insensitive. Even the slightest human emotion does not awaken in their hearts. They work in teams and routinely abuse their power and might.

Two meetings have been scheduled for today—one at 10:30, here in the office of the head of the British Museum, William Loudon, and the other at 3:00 in the afternoon with Sir Joseph Banks of the Royal Society.

As Fitton pushes the heavy wooden door ajar to step in, the large man seated at the table smiles officiously.

'Good morning, Dr Fitton. I am glad to see you again. There really was no need for you to take an appointment. We have met so often at different clubs and parties.'

'I've come to see you for a very special reason, Mr Loudon.'

'Please do take a seat and tell me all. I'm sure you won't refuse a cup of coffee.'

'Thank you, but that can wait.'

'My, my! It does seem to be a serious matter.'

'It could be serious depending on who I am speaking with. But I am not here for anything personal. Rather, it involves us as a community, as a people.'

'Community? You mean as British, the people of England? Please tell me clearly. Are we at war again? Waterloo has been over for a while now. Are the French at it again? I haven't heard anything yet.'

'Must it always be about war, Mr Loudon?'

'Why speak in riddles then, sir? Surely we both cannot be arguing for and against ourselves as British? Or have you come to pick up a fight?'

Fitton smiles. Maybe he should ease the tension first. If they continue like this, nothing will be achieved other than a fallout.

He says softly, 'There is nothing to worry about, Mr Loudon. I, too, have no interest in sparring with you. Please order that coffee now.'

Loudon rings for an orderly. His smile still in place, he says, 'Look, Dr Fitton, it may be fine to pick up an argument with a government official, but there will be no peace for sure. Because we are used to being blamed for many problems. No doubt we are quite at fault. After all, we work like sloths. But there is not much we can do. Our paths are usually jagged, and for every step forward, we have to move back two steps. We have no way out but to bide time. And then our hands are tied tight. How far and fast can we walk with our hands and means tied up?'

Fitton looks at Loudon calmly. 'Tell me Mr Loudon, do you think it possible for one man to prepare geological maps within fourteen years for the entire expanse from Devon to the northern parts of Scotland? To have walked across an entire country? To have mapped each and every nook and cranny of underground rock formation in land covering fifty thousand square miles throughout this region? To know each bit of land from the Welsh Marches to the shores of Kent?'

Loudon is suddenly a tad confounded, 'What exactly are you talking about?'

Fitton continues, direct yet calm, 'To do this takes a lot of patience. It takes sacrifice and the power of thousands of elephants along with the stubbornness of a bull. This takes immense vision as also a keen sense of how time must have unfolded in order to understand our four dimensions. One needs genius to be able to match and situate each such event that has occurred on earth eons ago, and to place them at the level of human imagination and logic. And that when there was no tutor to teach him, no wise elder to show the way forward. Now the essential question is, how could such a man and his work of such great significance go missing and unacknowledged by us? Or has England regressed to two hundred years ago from today and become like Venice?'

'Hold on a minute, Dr Fitton, who exactly are you talking about? Your eloquence reminds me of Shakespeare's poetic narratives. You have clearly come prepared, haven't you? How long did it take you to prepare this draft?'

Fitton smiles tightly, 'The draft was prepared last night. I couldn't sleep all night yesterday after my visit to the Fossil Gallery of your museum. I don't wish to waste my time any longer. So I prepared my speech for today in my head almost instantly.'

'Hmm.' Loudon's voice is gruff, and yet the smile remains pasted to his face. 'You visited the British Museum yesterday? And yet I have no information of it. Had I known in advance I could have shown you around. I could even have provided you with an excellent guide.'

'I hadn't come to simply visit the museum, Mr Loudon. I had come to make a stop at a special collection. I had to come as an ordinary visitor because you do not show that section to your special guests.'

'I see. Indeed, it is good to move incognito at times. But whose collection was it that has set you off on such an emotional journey, may I ask?'

'William Smith. Would you have heard this name somewhere?'

'Of course, I know the name. Strata Smith. A mad man. Wanders around collecting fossils. He was in need of money sometime back. Came begging to the museum for it. We bought his collection. In fact, a few established and well-known citizens had also put in a word for him. He was in a sorry state, the poor man. We were the first to buy his collection and bail him out.'

'Of course, you were compelled to buy him out. His collection worth more than fifteen hundred pounds was fixed at only five hundred pounds. Even that measly amount was not paid in full. And where have the fossils been kept, Mr Loudon, do you know?'

'Why? Didn't you see them in the showcases?'

'Is that true? Let's assume that you are not acquainted with the matter. Let me then recount it to you. This man, William Smith, delivered to the British Museum close to three thousand fossils, each washed and polished to a fault, all at his own cost. Some months later, he found out that these were all heaped together like trash in a damp airless room with no light. When he registered his displeasure with the authorities, he was told that there were no spare shelves. The fossils would remain as they were. He finally got the authorities to agree to a small remuneration with which he himself got down to building some shelves. He even paid to transport these to the museum on his own account. Then he and his nephew took the utmost care to arrange these fossils. They labelled each fossil all over again.'

'Excellent! You know it all. But it's a simple matter. The one who gives the fossils also gives the shelves. Smith was paid for his labour by the government. The matter was closed. That was it.'

'Have you ever taken a second look at these shelves, Mr Loudon?'

'I am the Director of this vast museum. Neither is it possible for me to check such shelves on a daily basis, nor has a schedule for this been drawn up for me. Men have been assigned to do this job.'

'Fine. Then consider this. Would you be able to open this collection for display tomorrow? Should there be a visit by a foreign delegation of scientists or, say, the King or Prime Minister of England themselves?'

Loudon does not utter a single word. But the smile remains frozen on his face. His expression says that he has been told a funny joke. He continues to fiddle with the ink pot on his table.

'So, this is your community matter? And I was thinking of all sorts of important concerns. Couldn't you find a better subject, instead of this mad man, Strata Smith? I am quite bewildered by your behaviour.'

Fitton enunciates, slowly and calmly, 'The condition of the showcases are pathetic. Most of the glass panes are broken. The room is infested with rats, covered in cobwebs, dirt and dust have taken a hold over everything. Is there really nothing for you to do about these issues, Mr Loudon?'

'You, I am sure, will understand, Dr Fitton, that I have no time or even means to do anything of the kind. There are so many such hobby collections here, presented by so many different Smiths. The annual grant that is given to us is not even sufficient to clean ten per cent of the museum's premises. What all can I possibly take care of, tell me that?'

'William Smith's collection is not a hobby collection, Mr Loudon. It is a work of science. It is the most brilliant documentation and study of the creation of the earth. Right from our ancient past till the present, of its evolution and transformation.'

'Please keep these studies and documents to yourself. Some of you, respected members of our society, get these sudden bouts of do-good-feeling for some random people, and the craving to set them up as people of eminence. And all the problems and blame land on our doorsteps, us government officials. I have heard the man had to serve prison, and then he ran away. Why don't you find him? Confer an award on him, set up lectures and conferences to honour him? Invite me, too. I'll certainly join the ceremonies. And perhaps we can then even think of getting the showcases repaired. Rest assured, Dr Fitton, no one will dare to throw out anything that finds its way into the Museum once. Smith's collection will not fall into the wrong hands, you have my word for it.'

'Perhaps not yet, but it will soon enough. The collection that is open today to rats and moles, will one day be confined to safety vaults, accessible only to some special members of the scientific community.'

'Bravo! I can't tell you how happy I feel listening to words of such positivity from people such as you. Such eloquence and insight. In truth, it is the best time of our lives when men can dream unbidden of changing the world. May that day arrive soon, amen. Do keep on with your efforts, Dr Fitton. Please do ask for all the help you may need. You need not hesitate at all.'

***

For the meeting that afternoon at 3, Sir Joseph Banks arrives late by five minutes. Immediately, he asks to be forgiven for the delay. And he says, even before Fitton can utter a word, 'Why are you getting so worked up about Smith, Doctor? You are meeting different people, continuously questioning them. How does this help you?'

Fitton takes a while before saying anything. He has immense respect for this wise old man. He answers, slowly, softly, 'Should I tell you the truth? I reveal my concerns because of my own personal need. I had been to visit Mr Cary to check out Smith's maps which are in his safekeeping. And what I was told left me completely dumbfounded. My curiosity about this amazing man has increased therefore. I have visited many. It is beyond comprehension how unjust and unfair people have been with him.'

Sir Joseph responds with immense calm. 'I know. You could call some of it a problem of his behaviour, but mostly his misfortune.'

'You call it misfortune? While a bunch of mediocre men have grabbed several prestigious positions simply through the weight of their wealth and lineage? They are so insecure about themselves that they fail to accord even basic respect to a deserving man. They have formed a tight group together. For some unknown reason, they fear certain hard-working achievers. They are perhaps afraid that if these men are recognised for their achievements, it may raise suspicion about their own capabilities.'

'This has been happening forever, Doctor. An atmosphere of competitiveness is building slowly in just about every field, be it in literature, art, or even sports. Qualified persons are prohibited from entering such groups freely. One has to try and adopt several approaches to find acceptance. Today's England is not the only witness to such incidents. Indeed, the world over, in every single country, this dirty game has been played across the ages. Even we, you and I, have perhaps taken advantage of such practices.'

'That we are guilty, I do not deny. And yet, I do come with hope to you and those like you, blessed with a generosity of spirit. At least for Mr Smith, do please try the best that you can.'

'Who has told you that I have not tried? But I acknowledge, yes, that I have not managed any headway in the matter. Perhaps, you are right. I haven't tried enough. I should not have refused the membership of the Geological Society, simply because I experienced a blow to my self-respect. I do not know what I am going through, but I do not find myself strong enough to oppose and take a forceful stand against things anymore.'

'This state of affairs cannot continue for too long, sir, it cannot. In the days to come, we will find hordes of new faces from different strata of society come forward to take their worthy positions amongst us. I do hope that they will be able to rise above all this dirtiness of class, position and lineage.'

'Yes, your hope instils hope in me. And yet I fear that the very opposite of this may come to be. Our future will be witness to much more foul behaviour. Till now, for us, community has only meant our lineage. But the future will see many more divisions—groups divided on the basis of established talent, creative talent, teams of those who have the ability to manipulate and fake talent, mediocrity that may establish itself with the support of some truly established names, and a class that will trade in talent as well. As time will go by, the unending desire to achieve fame will foment more and more insecurities. I do believe strongly that men like Smith, men whose only possession is real talent, will sink into more deplorable conditions.'

'Does that mean talent will not receive even the respect it deserves?'

'It may receive respect, but by then it will be too late. Those whose due it is, may not even be alive to witness that happy day. But try as I might I cannot be hopeful either that these conditions will impart a lesson to our future generations. Because the desire for fame is always so strong that this sort of sordidness will continue.'

'What you are saying only creates greater disappointment.'

'Why should this disappoint at all? Accept it as eternal truth. The more you attempt to build a class that is educated, this privileged mixed bag will resort to that much more dirt and muck.'

'Greenough, Farey, and such others are plagiarising Smith's various maps. They are publishing these under their own names. These new maps are being sold at a very low price and are riddled with errors for which these men have been reprimanded, in writing, by some members of the scientific community.'

'A copy of one such letter is available with me as well. Mrs Bennett of Wiltshire, the highly respected scientist, has complained that in Greenough's map there is no mention at all of the Crokerton clay layers, whereas Smith had specifically mentioned to her that they are very much there, just beneath the layer of green sand. Smith's work is extensively published in several small journals and magazines. And hence, students and locals are well aware of many of these details.'

'Even the sun's rays are not powerful enough to dry up an entire ocean built painfully drop by drop.'

There is an artful smile on Bank's lips, 'That's a good one, indeed, the brightness of the sun. Tell me more about what's being said about the sharpness of this sun's rays.'

'A few days after Greenough's map was published, a lady, Scientist Webster, ran it down as incomplete and incorrect. We need to tear down the web of lies people like Greenough are weaving. It's a tragic day when a man claims to be a geologist without ever going into a forest or climbing a mountain. How can one become a geologist with absolutely zero knowledge of minerals and lithology?'

'It's good to hear you say this. But it'll have to be more than words to have any effect. And who may I ask will bell the cat? Do you have the capacity to find Smith, put him on stage and confer an award on him?'

Fitton looks down. 'I confess I do not. But you, at least you, are not tied down by anything. Can the Royal Society not call for a meeting? It only needs your agreement.'

Joseph Banks guffaws.

'Do you think the President of the Royal Society is Napoleon Bonaparte? And that all the other members will bend down to agree to his every command? No one wishes to exit the circle of government diktats and officialese. And Greenough and Loudon represent that very rule. If they are dethroned, then we may see some positive changes.'

'Isn't that extremely surprising? A private club formed just twenty years back is now the most important arm one can think of. And they are hammering away, bit by bit, at all the values, rules and regulations of all the old institutions. They confer fame as they please, and ruin reputations as they wish. Even our worthies like Sir Joseph Banks and Reverend Townsend are fearful of them.'

'There is really nothing that I can do, Doctor. You have placed your hope on the wrong man.'

Fitton changes the topic. 'Have all of Smith's payments been cleared?'

'Well, a certain Mr Conolly, an estate agent, has filed a case. This is the man who fooled Smith into buying the Tucking Mill Estate as well as the golden limestone mine in Combe Hay town.'

'That is a very strange incident, don't you think? That Smith, the same man who helped scores of people procure some of the best mines across the country, should buy such a degraded mine for himself? Would he not have known that the quality of golden limestone in Combe Hay town is very poor?'

'He may have known. Maybe he got greedy at its low sale price. He wanted desperately to become a mine owner perhaps. Just like the chief engineer who dreams of building a palace for himself having built palaces all his life for others. After all, he had bought several properties, maybe five or six, and that too, when he had no fixed income. Although, it is true, that Mr Conolly managed to wrangle much more out of Smith than it was worth. He even auctioned many of Smith's possessions and invaluable minerals and crystals from his previous precious collection. Now, there is nothing, not even a house, to Smith's name.'

'Where is Smith right now, sir?'

Joseph Banks gives Fitton a cutting look, just for a moment. And then his eyes light up.

'What do you think? Am I hiding him?'

'Well, not exactly, but I do think you have tried to locate him.'

'And suppose I have found him, how would it change matters if I told you about it? Will you take him a bouquet of lilies? Or perhaps inform Greenough and party to nab him again?'

'What are you even suggesting, sir? I despise people like Greenough.'

'People forget what they really feel when they sit beside each other at special gatherings, wine glasses in hand, warm greetings on their lips. Do not take offence, Doctor, but people do not wish to be deprived of the pleasure of gossiping and criticising each other. Particularly against men like Smith, talented geniuses with nothing to protect them. It's not even too difficult to spread tales about their personal lives and shred their reputation to pieces.'

'Let that be then. I'll not enquire after his whereabouts anymore. You are right, what good would it do to know? But I am more concerned about Smith's nephew. I believe he is an extremely promising boy.'

'I've heard the same. Smith had him live in his house for some years, teaching him everything he could. But then he was in prison and the boy was thrown out of school. The principal at the school where he studied is a friend. He told me that the boy is impossibly brilliant. And he has gained a tremendous amount of knowledge having lived with Smith. He is of the opinion that the boy will go a long way and garner fame beyond belief.'

'Perhaps it is better not to speak of the future, sir. We have already done our best to destroy it. If the boy is so bright, could the school not have raised funds for him?'

'Even before this could be done, the boy had his name removed from the school rolls. His fee for three months was due and perhaps the other boys made fun of him. He too suddenly went missing along with Smith.'

'Total destruction of talent and a family. Greenough truly is more fiery than the sun's fire.'

Joseph Banks laughs loudly again. But the reason for this eludes Fitton. Then Banks switches topics completely.

'Would you know of a new museum being built in Scarborough in Yorkshire?'

'A new museum? Museum of what? Would you know?'

'Did you not read about it in yesterday's newspaper? A blueprint was also published alongside. Possibly a circular plan, at least as I gathered from the newspaper. Please wait, I will show it to you.'

Joseph Banks hands Fitton a day-old newspaper, opens its last page and points to a small article.

'Look, this looks like a lovely plan.'

'This is a rotunda pattern. Even London has only a few homes planned with this shape. It's a special structure called Doric. Looks like a high building.'

'Here, read this. It says that it will be about sixty feet high. And it will be several feet under the ground as well. The museum will be right at the shores of the North Sea.'

'And who is building it?'

'It's not a private property. But it is being built with support from Sir Johnston, the President of the Scarborough

Philosophical Society. The locals are perhaps donating generously as well.'

'And who is the Director?'

'I do not know too many details. But whoever it is, he has tremendous knowledge about geology. The museum will have shelves throughout each floor, from bottom to top, with notes for each fossil and their details arranged in each shelf. Each geological period will be explained in such a manner that the visitor will feel as if they are walking through an England that evolved through the ages from more than two hundred and fifty million years ago till the present.'

'And who will have those many fossils for display?'

'One needn't worry about that. At least fifty per cent of England's citizens have loved and collected fossils over the years. This museum will be made of such personal collections.'

'It will take a good ten years or so for such a huge museum to come up, I would think.'

'I think it will be ready in the next three or four years. There will be no dearth of funds.'

'That sounds wonderful. Please do let me know when it is ready and we could visit it together.'

'Yes, I would love to do that. But only if I'm still around. I sincerely hope that this new museum will dispel the despair that has set in you since your visit to the British Museum. But I must now bid you goodbye. I'll be called in for supper soon.'

Fitton walks to the door. Then he turns around again.

'May I take the sheet of paper with me, please?'

'Which sheet?'

'The one that mentions the Scarborough Museum.'

'Ah, that! Yes, do take it. It might offer you more insights.'

'Thank you, sir. And good night.'

***

Several thoughts course through Fitton's mind as he heads home, the clop of the horse carriage matching them. He unfolds the newspaper that he is clutching. He manages to read it again in the fading daylight. And when he finishes looking through, he sits back, silent and shocked.

He should have looked through the newspaper carefully when Sir Joseph was showing it to him. A rotunda blueprint is indeed printed here, but there are only a couple of lines of information. And not a word about its height, the fossils, it's funding, not a single other detail is mentioned. The work of building the museum has not yet started. How then was it possible for Banks to be informed about all these details? Now Fitton is beginning to feel uncomfortable. He should have paid greater attention to these details when the news came up. But what is it that he is missing? Why would Sir Joseph have suddenly shown him this Scarborough blueprint? Was it only because he's an interested geological scientist? That must be the reason, no doubt. And yet, something is bothering Fitton still. Is there something hidden behind Banks's words? 'I sincerely hope that this new museum will dispel the despair that has set in you since your visit to the British Museum.' Who is the Director of this museum? Sir Joseph gave out so many unpublished details about this new venture, and yet he couldn't say who its Director was. How could this even be possible?

Fitton shakes his head and chuckles. The more he digs into Smith's history, the more paranoid he himself is becoming. He must brush off the Smith narrative from his mind. It is not his duty to look for Smith. His purpose, rather, is to manage a look at Smith's many unpublished maps. And to do that, he must do what Cary has directed him to. And indeed Fitton is on that path.

Now, it is urgent that he visit Cary's office once again.

YEAR 1823, LONDON

Day after day in a dark damp town life eddies on like the wild winds. The chill of mountain ice seems to penetrate through skins, and through cold fur coats as rain pelts outside houses, forming pools across narrow lanes and splashes down umbrella ribs of the unfortunates caught in it. And then there are the muddy showers thrown up by carriage wheels as they rush by to the canter of horses, adding to the frosty nip in the air, as all eyes look up, waiting to devour even a weak sun should it deem to show up.

Someone walks out of Cary's office. Fitton senses that the man has pulled his umbrella down, just a little more, in a desperate attempt to hide himself. After waiting a few moments, Fitton looks back again. The man is clearly ageing, his walk is a shuffle. His agility is gone. He is dragging his left foot. Perhaps he is lame or even ailing. It's a strain for him to climb into the carriage. He moves a step up and then pulls his other foot forward to rest next to it. The left is certainly almost lifeless. He crawls up to his seat. Fitton catches a glimpse of his chin, his beard is salt and pepper. It's not been trimmed, left uncombed for days. A blue vein-riddled arm shoots out and bangs the carriage window shut.

There is no entry at all for today in Cary's visitors book. Perhaps the man is an old friend. After hesitating a while, Fitton questions the old peon sitting by the door, 'I don't see an entry here for the gentleman who just walked out?'

The peon is a grave man. He replies gruffly, 'He is an old friend.' Then he puts Fitton's name down on a slip and walks into Cary's chamber.

When Fitton walks in a little later, Cary is pulling on a pipe, staring at a map laid out on his table. He greets Fitton and invites him to take a seat.

Fitton begins, 'It seems the rain has not dampened the spirit of your buyers.'

Cary laughs, 'Buyers? Where?'

'I saw someone walk out just now, an aged gent, although there is no entry in your visitors book.'

Cary sticks the pipe back into his mouth and then folds the map on his table with great care.

'He's an old friend, drops by at times for a good chat.'

'Do you have the leisure for chats, Mr Cary? What I mean is, you always seem so busy.'

'One doesn't think about time when meeting old friends.'

'A friend? Is that true? Somehow, I thought he was a surveyor. And the open map here now confirms that it was work you talked about.'

Cary looks at Fitton, his eyes calm and unmoving. He blows out smoke. 'People in business chat differently. What else did you observe, Dr Fitton?'

Fitton replies, equally collected, 'Well, let's say the person found it difficult to walk. His left foot was giving him trouble. It's quite difficult to comprehend why such an ageing man would visit a map publisher in such blinding rain just for a chat.'

'People are driven by different needs, different attitudes. For example, you. It's amazing how you keep coming to look at a geologist's maps without his express permission.'

'Yes, I do not have his permission. But I come with the hope to receive it. I've met your conditions, Mr Cary.'

'I am aware. I've also heard that you have shaken up Loudon a little.'

'I do not have the capability to shake those men at all. They are fixed, wherever they are positioned, like statues of stone.'

'You speak very well, Dr Fitton. It seems Smith's matter has forced you to think differently.'

'Is that not expected? I find it difficult to believe how you can stay unaffected even after knowing of this from such close proximity. Do answer this, how many copies of the first edition of Mr Smith's set of sixteen maps were sold?'

'Around four hundred copies. All of these were sold out within the first week of publication.'

'I paid seven pounds for the set I purchased. My set was mounted.'

'I published three different types of sets. It cost five pounds for the ones that were only paper maps, unmounted. The mounted copies to be hung on walls were for seven pounds each. And the third, framed in leather-like-canvas, for twelve pounds each.'

'The maps published by the Geological Society are much cheaper, of course.'

Cary chuckles as he asks, 'Have you seen those maps?'

'Why wouldn't I? They are full of factual errors. While imitating Mr Smith's...'

'It's not imitation. It is called forgery. The fools can't even copy with any degree of command. Surveyor Farey professed to be Smith's friend. And Smith has never been an organised man. He couldn't even recall which maps he had handed over, and to whom. Farey traced many of his maps. But it astounds me how accommodating all of you have been. On the one hand is a man who worked himself to the bones for sixteen or more years to create these invaluable maps. And on the other is this spoilt grandson of a quack, who makes wonderfully detailed maps in just two years. And not a single question raised by any one of you to counter this.'

'You already know why questions couldn't be asked. But there was, no doubt about it, a strong need to know the truth. And you seem as if you have that answer. Why don't you tell me how?'

'Well, Greenough and his team collected the information by digging after Smith's work. They did not collect such data scientifically. You may have noticed how many educated aspirants specialise in this area of research nowadays, almost like intellectual spies.'

Fitton has no problem in grasping who is being referred to. And yet, he asks, all innocent, 'Is that true? How dangerous! What exactly did Greenough and company do?'

'Whatever Greenough and his apprentice did can't be called geology. Farey knew about all of Smith's haunts, wherever he worked, the fields, hotels, guest houses, even the rented homes in the villages, he knew of every single spot. They would visit these places. Greenough has enough money and power. In every hotel and boarding house, they would first pay to be rid of all other boarders. And then, they would line up, one at a time, everyone acquainted with Smith's work. Right from the hotel boy to Smith's coachmen, the horse carriage owners, even those who had helped drill underground, coolies and labour, everyone bought over for a few shillings for whatever information they would provide.'

'Fascinating!'

'Theft is always a fascinating subject, Dr Fitton. If the gatekeeper in my office buys pencils and papers and begins to make atlases on the basis of what he sees happening every day at our premises, what would you say? This is something quite similar. We are comparing maps here. A map that is eight feet nine inches by six feet, and which maps absolutely every layer under the ground of two countries. A map that has each available location of these regions, pinned down correctly. And all this, completed within two years only. And not one man's voice stands against this with an iota of doubt?'

'The matter is not as drastic as that, Mr Cary. I have heard Greenough's maps are not selling well at all. There are many voices within the scientific community which have run these down as merely tourist guide maps, with no geological substance to show.'

'The ordinary buyer does not understand these fine points, Dr Fitton. They only understand prices and recognise the government stamp. For such buyers, it is far more attractive to pick up cheaper versions of maps sanctioned by the Geological Society rather than spend money on something that a private agency has published.'

'I acknowledge that this is a sad state of affairs. And there is not much to be done at the moment.'

'You can do if you so wish to. Greenough's map is a fake, full of errors. Why do you not deny it publicly?'

'How much can I do alone? Society guidelines come into force through a process of voting. Greenough's team is extremely powerful. Can publishers such as you not come together to do something?'

'I am not up to standing against Longman Press, Dr Fitton. I must do business with many mapmakers other than Smith.'

'In that case there is no point in accusing each other of having failed Smith. For now, we are both unable and unwilling to protest against this injustice for our own reasons. Would you please show me Mr Smith's maps? Please do not turn me away again today.'

'Please follow me.'

***

It is after the sun has gone down that Cary knocks softly on the door to his map-room.

'It is time now to close the office, Dr Fitton. Please pack up in five minutes.'

Fitton comes out in that time. He goes to the basin and washes his face. After towelling himself, he sits down in an easy chair and shuts his eyes. Cary passes on a mug of coffee to him.

'I'm sorry that you're feeling exhausted. But I couldn't have given you more time than this.'

Fitton sips his coffee quietly.

Cary asks again, 'What do you think, Dr Fitton?'

Fitton raises his tired eyes away from the steaming cup and looks at Cary.

'Did you say something, Mr Cary?'

'What do you think of the maps? Will they work in the market?'

'I wouldn't know.'

'But I am a publisher. I must consider the economics of this carefully before I do something. Although he is quite an excellent cartographer, don't you think so, Dr Fitton?'

Fitton mumbles, and Cary can't catch on to his words. What's taken over this man? Had he fallen asleep while poring over the maps?

'Dr Fitton. I hope you are well?'

'Yes, I am fine. But I am extremely troubled. I am not able to express myself well. If we lose a man such as this, history will never ever forgive us, Mr Cary.'

'Who are you talking about? Perhaps, you should have been a writer, Dr Fitton. You are such an emotional man. What troubles you so, sir?'

'I am just an ordinary man. It is thanks to the advances in science that I too, like so many others, feel curious once again to observe the world around. I am happy to think of myself as a man of science. When people call me a scientist, I feel delighted. But as I looked through Smith's work, I realised that none of us understand the true import of the word scientist, or apply it as is true to its spirit. But, about its sale, I have no opinion to offer.'

'Hmm,' Cary is serious once again.

'Mr Cary, how much had you paid Mr Smith for his work for the first edition?'

Cary is taken aback. And then he begins to laugh. So loud and long that Fitton is uncomfortable.

'What is there to laugh about, Mr Cary?'

'Are you planning a biography on Smith? If yes, then let me tell you that someone is already at work on it,' Cary answers, still laughing.

'Who is writing Smith's biography?'

Cary finally quietens down. 'I was just joking. You ask too many questions, Doctor. What do you plan to do with so much information?'

'I am truly sorry, Mr Cary. I am asking for unnecessary details. In truth, everything inside me is downside-up after looking at Smith's work.'

'Twenty-five shillings. It was agreed that he would be paid twenty-five shillings for each set sold.'

'What? Just twenty-five shillings? No matter what the sets were priced at?'

'Yes, Dr Fitton. For all three types, whether the maps sold at five pounds, or seven or twelve. And he was happy with this arrangement. But I have still not given him his due on these four hundred sets. Let it stay with me. I'll hand it to him whenever he asks for it.'

'I see.'

'You do not think of me as an upright man, am I correct? I agree. I am not perhaps as moral as that. But you know as well as I do that he would have spent it all, no matter what amount I gave him. Perhaps he would use it for the mortgage of another property. This is a better arrangement, would you not agree?'

'Whatever happens, perhaps happens for the best.'

Dr Fitton sighs long and deep. His slim frame trembles, ever so slightly.

'Good night, Mr Cary,' he says, and picks up his hat from the coat hanger before he walks out of Cary's door.

YEAR 1824, CASTLEFORD

The Buckingham Street position has been gone now for the last nine months. Eight months and twenty-three days to be exact. When asked, Mrs Kitten's response always is, 'these last eight to nine months.' If one is too exact, Mrs Kitten is sure people will laugh at her, although they do proclaim loudly, 'Oh! I'm so sorry to hear about this.'

It is social decency to express sadness at other people's misfortunes although nothing so grave has happened that they should seem to express their condolence. In any case, this position had pushed her into a horrible spot. She couldn't let go of it, neither could she do her work well. Looking after a house whose owner has disappeared is like being trapped lifelong in an enclosed cell. To be honest, she was finding it impossible to continue there.

Last September, she had received a termination notice along with her salary. Mrs Kitten had not at all been disturbed after having read the letter. She had taken out her toiletries from a handbag and begun to get ready. Then she had put on the water to prepare coffee. And, finally, placed four ceramic coffee mugs on a silver serving tray along with a pot brimming with the hot liquid. She had carried these out of the front door and up to the main gate.

Two sentries had come running. That had been the first time she looked at them without getting angry. The rains would end soon, and the breeze still felt wet. Day after day, these poor men would continue to work at this dreadfully boring job, standing through hours of damp wetness. And to think of it, four sentries just to keep one man away! These boys may even have forgotten why they were sent here in the first place. They had never seen Smith, they did not have a photograph of his either. If Smith were to indeed turn up, maybe in disguise, armed with a false identity, they would not know him at all. It was simply to keep their salaries coming in, that these sorry lads were toiling away at this unwavering job for the last four years.

Mrs Kitten had smiled. A gesture of courtesy, but not artificial.

'Good afternoon. I got you some coffee.'

There were four mugs on the tray. The men had seen this, and they knew it was for them. And yet they had creased their brows.

'Coffee? For us?'

If only they could discover poison in their drinks, these men would have been thankful for this one eventful day in their lives perhaps.

The man who had narrowed his eyes had looked around shiftily.

Mrs Kitten had asked, 'Are you looking for my cat? Lucy? She is inside. Should I get her?'

He was embarrassed and had pressed his lips. The second man had said, 'Why would you want to get Lucy?'

'Because maybe you think I've poisoned the coffee. And you wish to test it on her first. That's why I offered. Why beat around the bush? In any case, you will need to look after her from tomorrow.'

The man had looked at Mrs Kitten. It was a long and cold stare. They had never had a conversation with her earlier. Every morning, this woman would simply march past them, military style, without looking in any direction. She walked straight into the house without fail between ten past ten and twenty past ten. And, every evening, she would march out again as soon as the gong went off exactly at five o' clock. Four years later today, what could be the reason for her unaccountable behaviour? Were they to stop guarding this house? But they had received no such notice yet.

The man extended his hand, 'The coffee is getting cold. Do give me a cup, please.'

'Here. And what is your name?'

'When on duty, I am W 28. "W" as in Watch. But at home, for my friends, I am Kevin.'

'That's a nice name. I am Mrs Daisy Kitten.'

'We know yours. This is Emon. Do give him a cup too.'

The man with the grave face smiled at last. 'Thank you for the coffee.'

'You are Irish, aren't you?'

'Yes. How did you know?'

'From your name,' Kevin had answered for Mrs Kitten. And then he asked, 'Should I call the others too? They are British. Nixon and Stan.'

'Of course. I'll leave the tray here. I'll pack up. I have to get back home early today.'

'Are you leaving, ma'am?' Kevin asked.

'You're a smart lad. It is my last day here today, yes.'

'May I ask you something? Has the restriction order against the person named Smith been lifted?'

Mrs Kitten had laughed. 'Not at all. I haven't yet received such bad news.'

'Bad news? Wasn't Mr Smith your employer?'

'I say so because that's how people like you will think of it. As for me, I've waited eagerly for him every day, and now I feel tired. I am beginning to lose my sense and understanding of everything good or bad that is related to him.'

'Where is your coffee, Mrs Kitten? At least have a cup with us before you leave.'

'That is a tempting proposition. But I don't have that kind of time. I need to put some things in order. I need to taste freedom again.'

'Of course, ma'am. Stan here could help you pack your things if you want.'

Mrs Kitten had shaken her head, laughing all the while.

Kevin had been perplexed. 'Are you thinking of something funny, ma'am? Share with us then.'

Mrs Kitten could not stop laughing. 'I'm sorry for my behaviour. Let me ask you a question. Do you enjoy investigating?'

'I like it, of course. But why do you ask?'

'Well, that's the profession you should have chosen.' Mrs Kitten had answered with a smile.

Kevin had smiled too, but the others had not been sure of what they were getting at.

'I'm sorry again, Kevin, but let me assure you that apart from a few personal belongings, I have nothing left here. And whatever I have, will fit into my tiny handbag. That's my tobacco pouch and four apples. So, I do not need help at all. May I take your leave then?'

Kevin had looked away, still undecided. But this woman had understood his motive. There was no point any more to keep the conversation going.

Exactly twelve minutes later, Mrs Kitten came out of the building again. She requested Kevin, 'Could you get me a horse carriage, please?'

Stan was sent for the task. Mrs Kitten said, 'Why don't you quickly check my bag while we wait?'

'There is no need,' answered Kevin. 'How much can you carry in a bag such as yours? I've been told that the man who had rented the house is in prison for unpaid debts. Surely, if he had gold or diamonds hidden somewhere, that wouldn't be the case.'

'That's correct…but…at least once…for the sake of duty.'

'It is not part of our duty to thumb your purse, Mrs Kitten.'

Mrs Kitten's insides had danced. What a laughable situation! Here were these boys for whom valuables only meant movables or jewels. About three months back, she had sensed that the job was slipping out of her hands. Her salary now never reached on time, and more often than not, she would receive letters, mentioning her tardiness or a complaint about some work that she had not carried out properly. She had been asked to explain her behaviour repeatedly. Everything indicated that something was about to unfold. The English are used to asking their employees to explain their side of the story before being asked to resign. Almost as if to create a scene of fairness, that the chance they had been given to prove their worth had gone abegging.

It hadn't taken much to see through this scenario. Mrs Kitten had seen through it, too. No matter what her explanation, she would be found wanting. So, she had not bothered to send replies. But she had done something else, secretly. Each evening, she had folded Smith's enormous maps, one at a time and smuggled them away in her purse, hidden under her food box. Even that day, two maps had been secreted away in the same manner.

Of course, even if the sentries had looked in her purse, they wouldn't know what they were. They wouldn't know what maps meant. She could easily have passed them off as her son's school work.

She had taken a seat in the carriage and waved out. They had waved back. Her eyes had been wet. She had pressed a handkerchief to her face and sat back, weeping silently. She would never come back to this house again. She had to control herself before getting home. An autistic child was waiting for her. He would become nervous seeing her distressed. He'd then stomp around the house, moaning away. At that moment, she had been an hour and ten minutes from home. She had had to keep the time in mind.

Four months after having lost her job, Mrs Kitten had left London. The amount she had received after selling off her London house was not much, but it helped her to buy a small cottage with land around it in Castleford village. She hadn't disliked London. It hadn't taken anything away from her. Smith had to abandon this bustling city because of his own mistakes. A man like Smith would never have fitted in with a fast-changing city like it. He was incompatible here.

Mrs Kitten was leaving for another reason. It was better for her to move to York. Around two hundred miles away from London, it took close to three days to travel there by carriage. She wanted her son treated at the York mental asylum. He was a fifteen-year-old, who couldn't speak, with his hearing impaired, and behaviour tricky to manage. She would find it difficult to get him a room there. The journey back and forth from London to see a doctor would be a good six days. She wouldn't have been able to pull it off.

***

Castleford is a small village in Yorkshire County. It takes only a day to travel to hospital and back home. Mrs Kitten hasn't found work for herself yet but is in talks with an affluent lawyer's office. She needs to fix it soon.

Today, someone from her old London days has come visiting, eight months and twenty-three days after she left her job. She knows the man well. He's behaving as if he happened to be wandering around when someone pointed her house to him. But he's holding a piece of paper in his hand, her address on it.

After their initial greetings, Mrs Kitten asks, 'Where did you get my address, Mr Farey?'

Farey smiles sweetly but doesn't offer a straight answer. 'If you are busy, I could come another time, Mrs Kitten.'

'But why? A busy man should not fritter away his time. Since you are here already, do tell me what it is about.'

'I don't see your husband about? Sorry, if it is too personal a question.'

'I don't mind, Mr Farey. My husband is in the army. I believe his troop is posted somewhere in Italy right now. I had written to him about this new house. I am quite certain that you have taken the address from my letter to him.'

'How would it help you to know where I got your address from? Especially when you did not leave a forwarding address with anyone in London. Perhaps, you wished to live in exile, unrecognised.'

'Not in exile, no. In truth there was no one in London I needed to leave my address with. But had Mr Smith been around, I would certainly have done so.'

'How do you manage your household? Have you a job?'

'Not yet. And my house runs on my husband's salary, whatever he manages to send us, whenever he can. It is just my son with me. I worked hard to put together money for my son's treatment. I need to take him to a doctor regularly. And then there are his medicines, nanny, and so on. That is why I am here. York would be too far from London for my needs.'

'Are you referring to the York Retreat? I believe it is exorbitant.'

'My son doesn't need admission there.'

'That's a relief. Do start his checkups at the earliest. Would you be able to do that if you received five hundred pounds for now?'

Mrs Kitten stands like a block of ice for a while. Then she smiles and asks, 'Would you like some coffee? We could talk while you have it.'

'There are very few who would offer a coffee to the uninvited. A million thanks, Mrs Kitten.'

Farey is extremely disturbed. Mrs Kitten is in her kitchen, preparing coffee for him. He is not able to stay still, he begins walking around the garden. Not walking, rather, he's striding around. This is a beautiful place, pretty and quiet. One cannot begin to imagine the bustle and stress of London while sitting here. Has Smith ever been here? Has he stepped into this house? Farey will find out soon enough. Now that the woman has agreed to a conversation, there is no doubt he will extract all the details.

'This is wonderful, a very different taste,' Farey comments as he sips from the mug.

'It's made with beans from France. Of course, water here tastes different from that in London.'

'Excellent. Many thanks for it.'

'My employer, Mr Smith, loved his tea. I had tried every possible kind of coffee with him, but he never took to it. But then he was not the type of man to say so straight to my face.'

'Oh! I'd offered coffee to him several times, and he never refused.'

'He wasn't the one to object to anyone, particularly those not close to him. He never even forced his employees, no matter what. He was immensely grateful for whatever came his way.'

'You revere him, do you not, Mrs Kitten?'

'Just as you do, Mr Farey. After all, you still follow his scent, even though he has been gone these last four years.'

'No, no, not exactly. How do I even know how to start pursuing him?'

'Perhaps it is not Mr Smith himself that you are after. It's his work, his research that you are looking for, no?'

'Mrs Kitten, perhaps you do not know as much as you claim to. Every present has its own past, a history. When that history is torn apart, then the present becomes like an endless night with no daybreak in sight. Even if one sees a dimness in the horizon, it is still dark. You cannot call it day.'

'Perhaps, you are right. But I have seen the light, however dim it may have been, between the two of you. Even in this dim darkness, I have that memory etched in my mind. And while we are talking, let me inform you that I do not have any of Mr Smith's maps in my possession.'

'You are a reasonably intelligent woman. You didn't waste any time in getting to the point. Then let me get to the point for my visit too. Give me those maps. We need it for the country. We need it for the sake of science.'

'Forget the country, Mr Farey. Even if you and people like you had thought about science, you would not have allowed a man such as Smith to be lost. That man was sent to prison for a negligible debt. When he was freed, he was not allowed to even visit his own home. After creating so much trouble, are you now claiming to work for the sake of science?'

'I cannot repeat everything about this matter to you all over again, Mrs Kitten. I just wish to reiterate, please trust me and hand over those maps. I promise we will not dishonour Smith in any way.'

'In my mind, the word trust carries a certain specific meaning. In any case, it is not about trust or mistrust at all. Simply, I do not have what you ask for.'

'Fine, then. I'll take your leave. Please keep this envelope.'

'I understand bribes, Mr Farey. What you need in exchange for this envelope, I cannot give you. Thank you, but please take that back.'

'The word exchange might have financial connotations, Mrs Kitten, and yet it is not so simple. I see now that I am a complete novice when it comes to explaining myself. I could not make Smith see this, and I have failed to make you understand it as well. That is fine. I'll take back these five hundred pounds. But let me tell you something. While on your visit to the York Retreat, one of these days, do take time out to check its Cell Number 201. They may not even allow you to go in. You would, therefore, do well to collect some information on the patient beforehand. Then, if you request the authorities, beg them even, they may let you in. The asylum is no longer run very strictly since the Quakers took over. You will remember the cell number, won't you? 201.'

YEAR 1824, LONDON

It's been five years now, since Smith has gone missing. The government has made no effort to look for him all this while. Even Smith's friends have not displayed any desire to do so. None of Smith's unpublished maps have found their way out of his publisher's doors till date. There never has been too much curiosity among England's public about the man and his work. Never before, not today. Visitors do not evince any curiosity to check out his collection at the British Museum other than a few in the scientific community who ask after his work for their own research.

Fitton has been haunting the British Museum for a very long time now. The room which houses Smith's different showcases is not kept locked. Its door remains half open always. He could check out the showcases if he wanted, but the area is so dark and depressing that he cannot bring himself to walk in. It seems pointless to make people see Smith's brilliance. How does he evoke interest in the crowds that do not wish in the slightest to plant a flag for him in their memories?

He remembers Smith acutely when something in the newspapers jogs his thoughts. Today, the newspapers are carrying an excellent piece of information. The Scarborough Fossil Museum was inaugurated yesterday. The half-moon Doric style museum seems to stand like a proud swan on the shores of the vast blue North Seas. Such a huge structure, ready within a year of its commencement! But what surprises Fitton is that he has not been invited to participate in the inaugural ceremony. And as for whether the Geological Society has received an invite or not, he has no clue.

Although Greenough no longer holds the President's chair at the Society, he remains in full command at that institution. And yet, the atmosphere seems to be changing. Though the current President, William Babington, is a Greenough man, he has brought in a couple of new members of his own initiative. Of these, one is a Cambridge–Trinity alumnus, Adam Sedgwick, and the other, his friend, Roderick Murchison. Both entrants are well-known and well-respected. One, a Cambridge professor, the other, a protege of Sir Humphry Davy. Affluent, blue-blooded, and extremely knowledgeable. These two care a fig about Greenough, and therein lies the problem. Is Greenough now being pushed to a corner?

In two months it will be time to vote for a new President of the Society. One can sense the whispers amongst the members already. On the pretext of showing him a map of the Woburn Estate, Farey drops in one day.

'What is your reading of the upcoming Society elections, Dr Fitton?'

Even though there's not an iota less of civility and politeness in Fitton's behaviour, Farey knows that this man does not like him much. In fact, Farey himself is completely disinterested in the results of the upcoming elections. The reason being that Greenough has not yet bothered to make him a member of the Society. Though Fitton thinks that the matter of who makes it as President of the Society will not make any considerable difference to Farey's life, it is not exactly true. Because, Farey still has some of Greenough's protection, which also allows him to nurse a fine ray of hope that even if the wait has been long, there may yet be a chance of his making it into this exalted group. And a member's position is all that Farey needs. Even if he makes it for a month, that will do. It'll help him lure clients when they see this particular designation mentioned just below his name.

Some people get pegged as loyalists. Farey too started as one such. He learnt some law to help in his father's business, but soon took a position with a duke's estate as its caretaker. And although the Duke introduced him to everyone as a surveyor, he never assigned any survey work to him. Once, on a hunt to find a surveyor-cum-engineer for his employer, Farey had bumped into Smith. He began cultivating a friendship with this odd man. Impressed by Smith's work, the Duke began thinking of buying mines. Farey couldn't even manage to map the layers under the mine, and so he was forced to stay as an assistant for a long time to learn some more from the blacksmith's son.

And then, he met Greenough. Finally, when Greenough's map had been edited enough to be published by Longman, Farey realised that his own name was nowhere to be found, neither in the preface, nor in the acknowledgements. He did not have the courage to ask Greenough to explain himself. In any case, he knew what the answer would be, even if he did ask. He would be dismissed for being neither a geologist nor a Society member. Greenough's reasoning: people would get suspicious if they saw Farey's name in print. And this too was why he didn't wish Farey to join the member's list. And yet these last few years, when Farey had been painstakingly tracing all of Smith's map by hand to give over to Greenough, he had heard the predictable litany of grateful thanks.

'Please do join us on the day of the presidential elections. Bellas Greenough personally invites you to join him,' Farey tells Fitton.

Fitton responds, 'I believe the elections will be done differently this time.'

'How so?'

'Well, the person opposing Greenough for the position of President is already receiving extremely positive responses from many members.'

'Who would you mean, Dr Fitton?'

'The renowned palaeontologist, as well as Professor-cum-Dean at Oxford College, William Buckland. Have you read his research papers? He claims that millions of years ago, almost close to two hundred and twenty-two million years before today, there were giant creatures walking the earth. Creatures called dinosaurs. Unbelievably large. He is researching them, in fact has found fossils which he is putting together to form its skeleton. He calls this the Megalosaurus. It's a frightening structure.'

'Pffft! All bunkum.'

'Science is not bunkum, Mr Farey. Buckland has provided enough evidence by giving exact details. His papers are available at the Society library. Do read them, please. But, oh, I am sorry. I had forgotten. Perhaps you do not have permission to use the library, am I correct?'

Farey's jaw hardens.

'How can you be so certain that Buckland's candidature will be fronted by the Chairing Committee, Dr Fitton?'

'I could ask you the same. How can you be so certain that it is Greenough whose name will come up, Mr Farey?'

'He has been President continuously for the last ten years. It is therefore an easy guess.'

'Not just a decade, but two. Because even if he served four terms of two years each as President, all the others who took up the post in the interim were his men, each running the Society according to his whims. And yet I feel there will be change. But, in the end, we can only guess, for, there is no right or wrong about it. The hope for possibility is different for you and for me.'

'My guess too will not affect the selection, Dr Fitton. I am, ultimately, an outsider.'

'That is true. And yet, as always, even today, you have come to speak especially on behalf of someone. That is no less a surprise. And you speak now possibly, as always, for the sake of science, or even for the sake of the country, am I not correct, Mr Farey?'

'The answer to this is your earlier answer. The word selfish carries within it a touch of the self. Selfishness is not for nothing.'

'Greenough reiterates these lines at least once at every election. And he never fails to mention the elections whenever we happen to meet at parties or at the Society get-togethers. But this time he has sent you as a special representative. Does he too fear change?'

'Isn't that quite natural, Dr Fitton? Just as you have heard Buckland's name crop up, so must have many others. But could I convey your assurance to Mr Greenough?'

'Greenough is an old candidate. We all know him. I will, of course, keep him in mind. One other thing, a new museum has come up in Scarborough. Would you know anything about it?'

'I have only read as much as the papers have printed.'

'Well, I am in the same boat. There is no architect mentioned. You are always in the know of such details. I thought I'd check with you.'

'The architect is a Richard Sharp who belongs to York. But he is not the one who made the design of the building. He simply carried out the brick and mortar work. I'll take your leave now.'

***

Mr Farey spends the entire day of the annual meet of the Society in the Duke of Woburn's office. The Duke questions him, 'Shouldn't you be with your master today?'

Farey smiles, and answers, 'You are my master, my employer.'

'Well, that's a new one I'm hearing, Farey.'

'Not new at all, my Duke. Greenough is a friend, and I have stood by him and helped him, nothing more.'

'But even so…'

'Since you will attend the meet, I am bound to stay informed. I needn't hang around the grounds there.'

'What do you say, should I vote for Greenough?'

'You must vote for the one you consider deserving.'

'What a transformation! Even till a few days back you were canvassing for Greenough by visiting several Society members.'

'That is not true. I have always maintained that one must vote for the deserving. Whenever I am seen, I am seen as a Greenough man. It is for people to think as they wish to.'

'Hmm, you are correct there, no doubt. Forget all this. I say, there is finally an opponent who will give Greenough a run for his money. So tell me now, who do I vote for?'

Farey smiles again, 'As I said, the deserving. It is only right to use the right of a vote very wisely.'

Later that evening, when Duke Woburn returns home, Farey is still at work in the office.

The Duke announces loudly, 'History has turned a new leaf today, dear Farey. There are three new members in the Society now, and two of them do not have any lineage to speak of. And the big news of the day! Today saw the fall of Greenough's rule. Buckland is the new President. He will not run the Society according to Greenough's diktats.'

Farey smiles, slightly, 'It will be as you desire, as you vote.'

The Duke adds, 'Does my vote alone decide the President? But the change has made many happy, in particular the young scientists. They do not believe in blood or class. They are the ones responsible for Buckland's ascent.'

‘But these that you mention are a very small number yet. Buckland couldn’t have won simply with their support.’

‘I agree.’ The Duke smiles archly, ‘Your whisper-support campaign died a clean death.’

‘Canvassing is never enough.’

‘Then what is? There was no money exchanged here.’

‘Not money. There is a different force at work here. It’s called influence. When that is on the rise, man begins to understand his place more clearly. But you cannot hold onto this power for long if the power itself is negligible. To hold on to it, one requires vision. Like many other inepts, Mr Greenough had no vision. And if he had any, he did not use it correctly to his advantage.’

‘I agree that there was much he did not do right. He was caught out for the fool he had shown himself to be in copying Smith’s maps and passing them off as his own.’

‘Only fools get caught. Those who don’t, are wily.’

‘The new group will bring in many changes. Just wait and watch. I believe Smith will be restored his deserving honour and glory. But where is that man? I wonder, is he dead?’

‘For a new man on the post, it is a quick opportunity to win popularity by honouring such lost memories.’

The Duke smiles and turns calmly to Farey. ‘You are a smart man, Farey. You are educated and intelligent. You have worked with maths and astrology, researched these in great detail as well. And yet…’

'Not everyone is fated for the respect that could be theirs. Take the case of Mr Smith. Could he ever get the honour that should have been his?'

'But you have always avoided hard work.'

'Is hard work enough? Then Smith should have been the President of the Society today. Sometimes it is not possible to bypass fate, no matter how well you apply intelligence as well as hard work.'

The Duke walks up to Farey and puts his hand on his shoulder. 'Do you regret what happened to Smith?'

'He is my teacher and he was proud of me.'

'I may not be aware of it all, but I do know a little. Abraham Reese approached you for work on his encyclopaedia. You had told him about Smith. But Reese had considered you more educated than Smith. He saw you, therefore, as far ahead of him. In any case, I do think that work on any encyclopaedia should be done by someone with education.'

'That day it did not matter what anyone thought. That day Smith misunderstood me completely.'

'I am sorry, very sorry about this, Farey.'

'It makes no difference anymore, my dear Duke. I am a black sheep in the educated circles.'

'Let it go now. I suddenly remembered something. I believe the new Scarborough Museum is excellently built. Should we visit it together?'

'Yes, we could. But let's not just visit the museum. They hold a lecture workshop there, a series of nine lectures, which costs a guinea each. If you are willing to spend it, let's both go and attend the lectures.'

'Go to Scarborough for some lectures? What are these lectures on? Who is the speaker?'

'It's a series on geology. And only fifteen shillings, if you are a woman, for the cause of women's education.'

'You have lost it completely, Farey. Why go to Scarborough when we can have London and Oxford?'

'One can go to the depths of the oceans in search of a teacher. It would be good to attend. One would meet old friends. And he teaches geology very well.'

'You speak in riddles these days. Who are you talking about?'

'It doesn't matter. You do not wish to learn new things it seems. We will speak of this later.'

'That is better. I am exhausted today. And oh! I completely forgot to mention. Greenough asked that you see him tomorrow.'

'How many times did he ask?'

'What do you mean?'

'I mean, how many times did he mention to you that I should meet him? Once, twice, or more?'

'What are you getting at? You're being extremely strange today. Normally, how often does a respected man ask for the same thing? He has called for you, and you will go.'

'Fine, I will.'

'I had planned to visit an estate in Liverpool with you tomorrow.'

'Fine then, what time?'

'But how will you join me? I must go alone.'

'You are my employer. You pay me a handsome salary each month. My work with you does not involve consoling the outgoing President of the Society. I will come here for work tomorrow morning.'

The Duke is taken aback. 'Well, I agree, but you will gain nothing by annoying Greenough over this.'

'There is no connection between these two words—Greenough and gain. You are tired. Please get some rest. I hope that the morning light tomorrow will see you refreshed. And, then, if you still bid me go see Greenough, I will. But not because Greenough has asked for me. Only because you have ordered me to. Good night, my dear Duke.'

YEAR 1824, YORK

In the last five months, this is Mrs Kitten's fourth visit here. Her son is undergoing treatment at the York Retreat, a mental asylum. The medicines prescribed for the boy can easily see them through for over a month. She could have come a fortnight later. But it will be Christmas in sixteen days and the boy's father would be home then. He will be with them for twenty-five days at most. The poor man has been away these last two years. She wouldn't like to bring her boy to the asylum during Mr Kitten's stay with them. And although it would really have been better to have visited about twelve or so days later, a strange incident had taken place the day before yesterday. And so she's been forced to visit the York Retreat now.

This visit has nothing to do with her son. The matter which has forced her hand is completely unexpected.

The Retreat has been set up in a refreshingly new way by a Yorkshire Quaker named William Tuke. Tuke has always been a member of the Quaker Committee. The Quakers believe that God lives within man's soul. And the soul is the ultimate. If the soul can be awakened, men will progress. England has previously witnessed severe inhuman tortures applied to patients in mental hospitals in the name of cure. These tortures included being manacled, painfully blistered with metal rods, beaten up with wooden sticks and being thrown into ice pits completely naked. Patients could never speak about these brutalities because they were never allowed to meet their friends or family, not even those who visited often and regularly. Hardly a few patients went back home completely cured.

Tuke's close friend, the respected Quaker community member Hannah Mills, had died while in treatment at the Retreat. The incident had triggered a lot of criticism. Relatives of several other patients began bombarding the asylum authorities with letters of complaints. Finally, Tuke took charge and began overseeing the responsibilities of the facilities. Since then, for twenty years or so now, the fundamental principles and practices of the asylum have undergone a complete overhaul. Families are now allowed to communicate with patients at least once a week. The number of patients is now ever-growing.

Mrs Kitten's son has nowadays begun to respond to calls and questions. Even if not completely cured, the improvement is noticeable. He smiles when he spots his mother, tries to call her 'ma'am' through broken growls. His father will perhaps break down when faced with these changes.

Even after she has confirmed all the medication and its dosage today, Mrs Kitten continues to sit. Had it been another day, she would return almost as soon as she had met the doctor. She would have fed her son on the cart ride back. But today she continues to wait, clasping her son to her chest.

There are two chairs and a table waiting, empty, at the asylum veranda. One of the chairs is usually occupied by Mrs Simon, a staff member at the Retreat. Mrs Simon seems forever busy. She's always scribbling away in a huge register, filling up page after page. This is the asylum record register. If only Mrs Kitten could take a peep into the register and check the name of the patient in Cell Number 201. But that's an impossible wish because even when Mrs Simon goes to the lavatory, she locks the register inside a drawer.

Sitting now in the chair next to Mrs Simon is Sister Rowena. She too is making notes in a file. This woman, in her mid-fifties, is a Catholic nun. The patients here love her. Mrs Kitten's son calls her Ro. The boy will enter his doctor's chamber only when Ro holds his hand and walks him in. Ro unfailingly makes the sign of a cross and says a prayer to send off her patients home after they have met their doctors. She has unaccountable patience. She is the only one here who Mrs Kitten can think of for her request.

The incident had taken place the day before yesterday. A young boy of around twenty had opened the gate to her house and marched in unasked. Mrs Kitten was busy pruning her peonies. The boy had walked up straight to her. He didn't even have a beard yet. He was clearly exhausted, dried leaves entangled in his hair. He may have walked here on foot through the jungle path.

Mrs Kitten was not one to be easily confounded. A woman who walks through life alone with an ailing child does not have the leisure to be frightened easily. She had snapped, 'Yes, who are you?'

'You will not know me by name, and if I tell you about myself, you may not believe me.'

The boy's voice was far more polite and calmer than his stride may have suggested.

Mrs Kitten had countered, 'Tell me then why you are here.'

'You possess several of Williams Smith's maps. Please give them to me.'

The boy hadn't beaten around the bush. Mrs Kitten was amused.

'Are you in a rush? You made your point right away.'

'The real point must always be made right at the start.'
'That is true. But never as quick as this, my boy. Who told you that the maps are with me?'

'I know. You have them. Had they been in the old house, those people would certainly have published them. If not with Smith's name, surely with their own.'

'So you are not with the Greenough gang! Which gang are you with then?'

'I am not with any gang. Please give me the maps, ma'am, I need them.'

'You've been saying one thing or the other since you have come. First off, I do not have the maps. And even if I had them, why would I hand them over to you? I do not know you.'

'I beg of you, ma'am please do not use the maps for trade. Please return them to the one they belong to.'

'Look here, boy, you're crossing your limits now. I'll shout for help.'

'Call them then. You call people, and they bring in the police. Then you'll send me to prison. That is all you people know.' The boy's voice was choking.

'Don't you dare start crying. First, you march in unannounced, and now you think you can cry. Tell me, who has told you that I have some of the maps?'

The boy tried to control himself.

'The one whose maps these are.'

'What do you mean? Has Smith told you? Where is he?'

'I can't reveal anything to you.'

'Are you his son? But how can that be? He was childless till five years back.'

'He is childless still. I am a relative.'

'I'd been with him for fifteen years. I never saw any relatives. I didn't even see his wife.'

'But they exist, even if you never saw them. William Smith is my uncle, my mother's brother.'

'Yes, he had mentioned a sister. But she passed on long back. And I do not remember if she had any children.'

The boy had smiled in a knowing manner, 'That is why I had said you wouldn't trust me, even if I introduced myself.'

'How can we trust anyone today, tell me? Many have come before you, looking for the maps. They had even offered me five hundred pounds. But I haven't heard you yet and how much money you can give me.'

The boy had looked down, gazing mutely.

'Tell me, how much do you offer? I could give you a discount. Your people are smart, sending a young boy to me. It might awaken my maternal love, no?'

The boy's eyes had welled up. 'I can't even give you five pounds. I've had to stop my studies at college because of the expense. And whatever we earn, goes into my aunt's treatment. I'll take your leave then.'

'Stop. Wait. If you are Smith's nephew, tell me where Smith is. I will go and meet him personally.'

'That I cannot do, ma'am. I have promised, never to tell anyone. About his whereabouts. But this I can share, take time out and visit the York Retreat. The patient's name is Mary Ann Smith. Cell Number 201. Her treatment costs a lot of money. It would help to find the maps. But I'll take your leave now. It's a long way back by foot for me.'

The boy strode off, out of the gate, without giving Mrs Kitten a chance to say anything else.

***

Mrs Kitten helps her son get on his feet. Then she leaves her place on the bench. Sister Rowena is at the table, working, her back to them. Mrs Kitten's son walks up to her and pulls at her frock.

'Ro!'

Sister Rowena turns around, a large smile filling up her face.

'Oh dear! Haven't you gone back home yet, my darling?'

The boy chuckles, 'Ro.'

'Come again around Christmas. We'll be celebrating. You'll love it.'

'Sister, I wanted to speak with you,' Mrs Kitten whispers.

'Of course, tell me.'

'It's about Cell Number 201. A certain Mary Ann. I've been hoping to meet her for long. Would it be possible today?'

Sister Ro turns to Mrs Simon.

'Could you check the patient's name for 201?'

'Give me a few seconds to complete this tally.'

'It'll take just a moment. Please check.'

Mrs Simon seems slightly irritated, but says, 'It's Mrs Smith.'

'I am sorry, Mrs Kitten,' Rowena says, 'it's not the same patient I am afraid.'

And almost immediately, she says again, 'No, wait, I remember. It's Mary Ann Smith. Am I not correct, Mrs Simon?'

Mrs Simon seems even more displeased.

'The name doesn't matter. You need to take an appointment to meet patients here. And in today's roster, there is no visitor mentioned for Cell 201.'

'Please, let me meet her just this once. I'll take a few minutes only. William Smith is my brother.'

'Is this patient very violent, Mrs Simon?' Rowena queries again.

'How can I comment, Sister, I'm not a doctor, am I?'

'Who is Mary Ann's doctor? Please give me the name from the register.'

'What a bother, and in the middle of so much work. It's fifteen minutes to 1:00 and at 1:00 sharp, I leave for lunch,' Simon barks. 'I hope you'll manage to complete all the pending entries, Sister?'

'The page is open. Just read out the doctor's name.'

'Dr Paxon.'

'That's a blessing. I know Dr Paxon well,' Sister Rowena assures Mrs Kitten. 'But you must wait. Visiting hours begin at 3:00.'

'I have to travel far with this boy.'

'I am sorry, but that's the rule. And if you wish to come again, you must take an appointment. For today, I could check with Dr Paxon if he agrees.'

'But will he agree if you ask, Sister?'

'I believe he will. But you must go in alone, not with your boy.'

'Let me feed him then. But where would I leave him?'

Mrs Simon nearly explodes, 'Only five minutes to 1:00, Sister, I am closing up.'

'I'm sorry,' mumbles Mrs Kitten, 'I've delayed you inordinately.'

'Oh, don't worry. This is an everyday affair.'

Sister Ro is still smiling pleasantly. On a face such as hers, only two emotions can be at play. Either happiness, or sadness. And both, genuine.

A few minutes before 3:00, Mrs Kitten's name is called out. Sister Rowena asks, 'Where is your boy?'

'I asked the coachman to look after him.'

'Dear Lord, how foolish! There is a chill breeze outside. He'll fall ill. Go on, get him here quickly.'

'Can I take him inside?'

'Of course not, but it's not good outside for him either. It's much better to seat him next to his dear aunt Simon. She makes the most wonderful Dutch cocoa balls. We are all devotees of it. There are a few still left, aren't there, dear aunt? Come on, get them now, you've got a new admirer, Mrs Simon.'

Mrs Simon smiles, finally, indulgently. 'You are being very naughty today, Sister. I may need to complain about you.'

'Cell number 201. The one with the metal gate, the patient was brought in four months ago. She was displaying very violent behaviour. She was incoherent, always squatting with her head in her hands, and smashing whatever her hands could find. And although she did not have a tendency to hurt people, the possibility to do so was always there. The family was not allowed to visit her for the first two and a half months. They come now, but once a month. Now her aggression is no longer visible, and she is incoherent only once in a while. But when her headaches are extreme, she cannot recognise a soul. She cannot remember even her own name…' Sister Rowena gives Mrs Kitten a brief history of the patient as they walk briskly through the corridors of the asylum.

When they are in front of 201, the Sister lifts three fingers, to indicate the number of minutes Mrs Kitten has been assigned for her visit. She herself goes into a cell next door.

Mrs Kitten clutches at the metal bars on the door of Cell 201.

She calls out, 'Mary Ann, are you sleeping?'

The cell is completely dark. Only a grating sound emerges from within. Is someone moving inside? A strong stench of medicines reaches her nose. A bony hand grabs a metal rod on the other side. A dim face can be seen, but it is just that, an unrecognisable blur.

'How are you, Mary Ann?'

A feeble voice answers, 'I am having a tremendous headache.'

'Is it still hurting?'

'No. But I'm sleepy.'

'Does your husband visit you?'

There is no answer.

'Mr Smith. Does he come, Mary Ann?'

'I don't know. Perhaps he does. He feeds me milk, caresses my head. Is he here now?'

'No, but he will come soon.'

'Are you a doctor? Have you come to give me medicine?'

'I am your husband's friend. He's a very good man.'

'He is. A good man. But he doesn't come. He doesn't even take me away from here.'

'Time up, Mrs Kitten.'

It's Sister Rowena.

'Please wait outside now. I'll put her to bed. Please wait by the stairs. I'll join you shortly.'

As they walk down the stairs, Mrs Kitten says, 'You've done me a huge favour, Sister. I feel like weeping.'

'Did the patient speak with you?'

'Yes, she did. Another request, please Sister. Please tell me if William Smith has ever visited her.'

'I can say no more, Mrs Kitten. I cannot give out more secrets. I…only on trust…'

Mrs Kitten does not let her finish. She clasps Sister Ro's hand, 'Please, I beg of you, dear Sister. This is my last question. I will ask you no more. I swear in the name of God, I will not get you into trouble.'

Sister Rowena lets out a deep sigh.

'No William Smith has ever visited her. Another person, only him. That too, twice. The register mentions him as the nephew of the patient's husband.'

'May God fulfil your desire to help humanity, Sister. I will leave a packet with you when I visit next. Please hand over the packet to this same visitor when he comes again. Do not say anything to him, just hand the packet to him please. Would you do this for me, Sister?'

'Let me check the register if it mentions a date. But what if it is tomorrow or the day after?'

'If it is the day after, I will come by. But if tomorrow, I cannot.'

'Better still that you hand it to him yourself.'

'That I cannot do. Please do not ask me why. Just do this for me, please. You are a servant of God, you do not fear public shame. I have my reasons, please help me.'

'Fine. Let me check again. Let me see when he is due next.'

Mrs Kitten visits the asylum once again after twenty-one days. She has left her son with her husband at home. Sister Rowena asks, as she takes the packet, 'What if he asks me something?'

'He will not. My name is written on the packet. He will not say anything when he sees my name. May God bless you, dear Sister.'

YEAR 1830, LONDON

Two years ago, just before Christmas, a doctor named William Hyde Wollaston passed away. A renowned physician, he was sixty-two when he died. Large numbers of people had been treated by him for their kidney and prostate ailments, and an equally large number had recovered. It was, therefore, but natural that locals would be extremely upset at his sudden demise.

But, two diametrically opposed reactions could be seen when the news of the good doctor's passing away reached the scientific community. At first, as was expected, one witnessed grief and sadness. Not only because he was a well-known figure, but also because he had been an unparalleled observer. Even the tiniest of tiny flower could not escape his eyes as he cantered past jungles on his way to visit his patients. But those with such sharp powers of observation sometimes unknowingly cause certain incidents to take shape. Just as this doctor had done.

One day, he noticed a crack in the wall mirror as he was shaving. He began to study the crack in greater detail, and suddenly a few ideas flashed in his head. Very soon, he built a camera, the *camera lucida*. While experimenting in his laboratory with some heavy elements, he happened to discover a metal, palladium. While refining radium and writing down the process, he discovered rhodium, and also perfected a way to process platinum ore. He also invented an instrument called the reflecting goniometer which could measure with great precision the angles between crystals faces. During his lifetime, Dr Wollaston had written at least fifty-six research papers on subjects as diverse as pathology, physiology, chemistry, optics, crystallography, astronomy, electricity, mechanics and botany.

Now, let us proceed to the other equally momentous happening. A small item published in a recent newspaper astounded everyone who read it. The same doctor, who had done no research on geology while alive, or indeed written anything on the subject, had left behind a will which stated that he had bought some stocks worth one thousand pounds, and the yearly interest it accrued, should be used to present an annual medal to the finest researcher in that very area of study. As per the wishes of the donor, the medal should be made of platinum. But should the government not give permission for the use of platinum, it being one of the rarest and costliest of metals, gold could be used instead.

This was like throwing a spanner into the works. Geology was an area of science that the common man was hardly aware of. And as for countries around the world, only a few had any knowledge of it. This was the subject for which our respected and reputed doctor had left behind instructions for the award of a medal! A subject on which he had not even written a small note. To add to it was the question: Who can truly be defined as a geologist? In this country, in England, those who had gained repute as geologists so far were mostly people with large collections of rocks and stones, or those who had built museums. Or else, had collected fossils to beautify their homes. How could any of this be connected to science? And how could a rigorous study be done of this subject?

Very soon, Oxford will also hold a convention to debate these same issues. Two years ago almost to this day, in December of the year 1828, Dr Wollaston had passed away. The news of his will and its condition was published in January of the next year. The medal is designed and almost ready. In the coming days, the awardee's name will be announced. The first recipient of this Wollaston Medal will be selected without any political interference.

As for the deserving candidate, the selection committee will put them through several stringent processes, what we know today in our modern world as screening.

At least for this very first award winner, the selection must, and absolutely must, be free from any influence and completely fair.

Prof Sedgwick, the current President, and his protege Roderick, have been invited as representatives of the Royal Society.

Sedgwick chuckles as he reads the invite, commenting lightly to Roderick, 'Look, Roddy, look at who they have elected Chairman of the Medal Committee!'

Roderick is a gentleman, a baronet. Perhaps, this man sitting across from him is the only one anywhere to call him by his nickname. There is more conflict between this teacher-student pair, than there is agreement. They are more often than not at loggerheads. In fact, how these two have come together professionally is a matter of bafflement to all those that have been acquainted with them.

Baronet Roderick picks up the letter, 'Dear God! This mad man?'

'Mad? Who are you, my dear, to call the Dean of Oxford mad?'

Sedgwick is smiling widely. He will never give up an opportunity to banter with Roderick at any cost.

'You know nothing about this man then. He devours anything that moves. He serves rats, bisons, crocodiles to his breakfast invitees,' Roderick rattles off.

'Why shouldn't he? He may be an omnivore,' counters Sedgwick. He is enjoying Roderick's nervous energy.

'Of course, yes. And he himself tries out earthworms, flies, even mud!'

'True scientific spirit. My respect for him is growing by the minute,' Sedgwick provokes further.

'And listen to this. Your respect will now touch the skies. The gent is extremely suspicious of anything to do with Catholics. There is a rock, with blackish patches on it, in front of a certain church in Italy. People believe it to be the dried blood of a holy figure. Our respected Dean visits the place, licks the patches on the rock, and calmly declares it to be bat piss!'

'What?'

Sedgwick's entire frame is collapsing with laughter. 'I haven't met him in long. And now I am dying to meet him again.'

'That is but natural. You will be like two peas in a pod,' Roderick guffaws, too, no longer able to keep a straight face.

'Who is the more insane of us, Roddy, Buckland or I?' Sedgwick jests.

'Is this a choose-the-best-madman competition?'

'And if it were? Who then?'

'When is the competition to be held? On the first day of the Wollaston Medal Selection Convention?'

'Yes, don't you think it would be a blessed day?' Sedgwick questions seriously.

'Well, I have accepted you as my teacher, no doubt, but in no way, does it mean that I'll follow you blindly.'

'What am I even asking you to do? Do not do it blindly, but do it with thought and your eyes open. Fossils of certain amphibians had been found once in one of the layers of soil in Wales. Had you not gone there and stayed still, as unmoving as a tortoise, when I had asked you? And then I had sprinkled you with a fine layer of clay,' Sedgwick jokes again.

'But that was for the sake of science,' Roderick counters, trying to look grave. 'You had asked to see how a whole creature could eventually turn into a fossil.'

'And why would you consider this a project not worth your while? This too is for the sake of science.'

'I think we are going only as representatives, you as the President of the Society, I as its member.'

'But you are also a baronet,' Sedgwick cuts in.

Roderick says, 'Well, that is true. But the Convention is not the appropriate place to witness a fight amongst madmen. I wish to remind you of that.'

'You speak as if we are going to see two mad men fight it out inside an asylum. I meant only to ace up against Buckland,' Sedgwick pretends to be annoyed.

'Must one win by becoming mad?'

'Yes, one can only stump an insane being by catching him by surprise. I must do something very different. Stay prepared, we will leave for the Convention together,' Sedgwick signs off.[1]

***

'The Bible states that our Lord God created the earth in six days. Who amongst those present here today believes strongly in this statement?'

The Wollaston Medal Committee meeting has started with Chairman Buckland's speech. The speaker has begun by aiming an unexpected and unnatural question at his curious onlookers and now stares at them, waiting for a response.

Roderick turns to his friend sitting on his side and whispers, 'Do you see how the madness has been flagged off? What was the need to start with such a query in a meeting like this?'

Sedgwick responds, 'Have some patience, let's see which way the water flows.'

---

1 The Wollaston Medal is a scientific award for Geology, the highest award granted by the Geological Society of London. The medal, named after William Hyde Wollaston, was first awarded in 1831. It was originally made of gold (1831–1845), then palladium (1846–1860; this was a metal discovered by Wollaston), again in gold (1861–1929) and then palladium (1930–present). This medal, instituted long before the Nobel Prize, is equally prestigious and renowned.

'I hear silence all around. If you believe strongly, nothing can shake that believe, can it? Dr Sedgwick, how about you?'

Sedgwick answers softly, 'There's a lot to say, Sir. But I would like to hear you out first. Why don't you finish, Dr Buckland?'

'It's been fifty years or so since a certain region of Asia has been attached to us as a British colony. A country called India. Some sixteen years ago, a museum was built there. A friend of mine, as well as several senior scholars, had the opportunity to work on the language of that land. Like almost all other primeval communities across the world, the people there too worship several gods. But what surprises one immensely is that in their original ancient text, the divine is mentioned as a single entity. The rest of the Gods are counterparts of this single form, even different forms of nature. And so the ability to know the workings of this form has not been fully understood, even in this age of science.'

'How is this connected to the Bible?' counters Roderick slightly irritated.

'I am coming to that shortly. The Indians have a God named Brahma, the God of creation. The Indians call that moment of time, as long as it takes Brahma to blink, a *pal*, almost like a flash.'

'Just look at this man's temerity. Now he's building narratives around Paul, John, Joseph...' Roderick grumbles.

'Pfft! This "pal" is not the Saint Paul you all know,' Sedgwick offers lightly.

'It seems that the Baronet Roderick has something to contribute?' Buckland is quick to have noticed Roderick's discomfort.

'No, Sir, I have nothing to offer,' Roderick says.

'How many hours does it take to make a day, Mr Roderick?' Buckland continues to stare at Roderick.

Roderick's expression is hardening, 'What do you mean?'

'It means as it should! I am simply asking how long are our days, not our nights.'

'Well, at least from what I know, it is twelve hours. But if you have discovered a new time frame, I would not know of it.'

Sedgwick motions to Roderick with his eyes. Roderick, about to add something to his excited response, restrains himself.

'You are absolutely correct, Mr Roderick. This is man's calculation of time. According to ancient Indian texts, in the kingdom of Brahma, the Creator, a day is called a *kalpa*. A *kalpa* in Brahma's realm is 4.32 billion years in the human world. The *kalpa* is then divided into 14 segments, each of which is called a *manvantara*. Each of these segments is almost 306,720,000 human years. When each such *manvantara* segment is further divided into 71 parts, these sub-parts are called *divyayug*. Each *divyayug* is 4,32,000 years.'[2]

Roderick whispers again, 'How long will this lunatic ramble on for? Someone should stop him.'

Although he can't hear him, Buckland revels in Roderick's facial expressions. Impatience is clearly written all over it.

Buckland continues, 'Now the question is: how is all this relevant? I would posit, it is in the calculation of time. The Bible is an ancient text as well. There we read that the world was created in six days. But those six days cannot be calculated, or understood, according to how we, humans, consider time. Just as we see the value of different currencies change—let's say between the pound, dollar, dinar—there is a difference in the structure of time as seen in the kingdom of God and the kingdom of man. So is it six days or in truth, six aeons?'

'I do not understand a word of what you have said, Prof Buckland.' Roderick says now, his expressions grim.

---

2 The word *pal* translates to 'moment' in English. It indicates a brief period of time, or an instant. It can also be used to signify a fleeting experience or a significant moment in time. *Kalpa*, in Sanskrit, is literally formation or creation. In this context, it means a long period of time, or an aeon. In Hinduism, a *kalpa* is equal to 4.32 billion years, a day of Brahma (12-hour day proper) or one thousand *divyayug* or *mahayug*, measuring the duration of the world. Each *kalpa* is divided into 14 *manavantara* periods, each of which lasts 71 *yug* cycles of 306,720,000 years.

'Logic and research have provided you with an answer, Baronet. After all, Prof Sedgwick has shown us an age when fish came to inhabit the earth. The two of you have also presented evidence of the next stage when the amphibians emerged. I believe you are now working on a subsequent period as well.'

'All these events took place millions of years ago.'

'Exactly, Mr Roderick. Why not calculate how long a time it would be for a day according to the *kalpa* segment?'

'Astounding!' Sedgwick claps loudly. 'Reconciling religion with science in this manner is truly incomparable.'

'It is not my doing, Professor,' Buckland continues in an even voice. 'Had my old friend Nathaniel Wallace not made me privy to the ancient philosophical concepts of the East, such a thought would never have come to my mind. It is because we disrespect and, therefore neglect these ancient traditions, we are completely apathetic to their beliefs and culture. And yet, the observation powers of these ancients were much sharper than ours. They turned their observations into tales so that every one of them could understand these better. We laugh about the basic principles because we do not pay adequate attention to their observations.'

'Absolutely, sir! Let me tell you further that there was another age, prior to the fish, the age of soft marine creatures, the trilobites. Their impressions have been found on sedimentary rocks,' Sedgwick adds.

'Is this your research as well? If possible, may I see it?' Buckland asks.

'Of course. And let me add too, that evidence of six more ages have been found, each such age captured in the fossils found in the many layers of rocks. If only we were to see all of England, South Wales, and Scotland as one, a complete picture of all the ages will emerge as clear as a photograph. You are so correct in what you have said,' Sedgwick concludes.

'Have you done this work? How amazing! In which case the medal must be yours,' Buckland says excitedly.

'That time is yet to come, Prof Buckland. And, even if Baronet Roderick and I have played a large role in this, it is really a boy who possesses the facts of the six ages with factual evidence.'

'A boy! What do you mean? Is he a student or a young professor? Such complex research would be beyond an ordinary person.'

'How he has carried on his research would be best understood if he himself were to talk about it. He is thirty years old and wanders around day and night across the land. He has been working like this for the last twenty years. Would you like him called in?'

'Is he here? In this room?'

'I have kept him waiting at the reception.'

'We are digressing,' Baronet Roderick stands up. 'Let me remind the gathering that we are here to select a deserving candidate for the Wollaston Medal Committee.'

'Leave that be, Sir. Such tremendous work has happened and we are ignorant of it. We will make the selection carefully. If it is fitting, we could choose this boy as its candidate. Prof Sedgwick, please do ask him to join us.'

Roderick sits down, his face red. Sedgwick whispers to him on his way out, 'Do not be angry, Roddy, Buckland is a little unhinged. In his excitement, he has forgotten that you are a Baronet after all. Wait and see who I fetch.'

Within a couple of minutes, Sedgwick walks back in, a young man following. Thirty or so, his beard stale and unkempt, his clothes frayed. He seems to have borrowed whatever he is wearing.

Buckland does not wait for an introduction. 'All that can come later. First, tell me about the six ages and how they follow each other. Answer as I ask you. And if you cannot answer properly, do not wait to be asked to leave the room. Do you agree?'

The young man nods.

Roderick tenses, clenches his palms and holds them tight.

Sedgwick whispers, his tone reassuring, 'Do not worry, Roddy, he will do just fine.'

The youth unfolds a huge map and pins it up on the wall. And next to it, another with a layout of each layer of rock under the earth. On the table he places, one by one, hand painted pictures. Pictures of fossils.

And then he begins to speak.

YEAR 1832, OXFORD

The concluding meeting of the Wollaston Medal Committee was held in the month of January last year. The Chairman of the committee was Professor Sedgwick. The awardee's name had not been announced then. It will be done today.

In the 1st century BCE, in Rome, a D-shaped theatre called the Marsellas had been constructed. Inspired by the Marsellas model, but with a touch of the Gothic, a theatre complex, The Sheldonian, was built in the mid-17th century on Broad Street of Oxford within the university precinct. The then Chancellor of Oxford, Gilbert Sheldon, had donated one thousand pounds from his personal savings to begin the construction of the hall. And that is why the theatre was named after him.

Fitton is a little delayed in getting to the theatre on the day of the award ceremony. Coincidentally, he bumps into Farey at the complex gate. Farey comes forward to greet him.

Fitton apologises, 'I am sorry to have been delayed.'

'I am in a similar situation,' answers Farey. 'I'd appointed a coach last night, having agreed to twice the fare. But it was delayed because one of the horses fell ill. However, by God's grace, here I am.'

'It's quite crowded. Let's see if we find a seat inside,' adds Fitton.

'Yes, let's. But of course, there will be a seat reserved for you. After all, you are a member of the Society. I am an ordinary onlooker. I will sit in the aisle,' says Farey.

'Come sit with me if there is place. We've met after long. At least we can have a conversation.'

'I am grateful. You are no longer angry with me, I hope,' Farey smiles.

'Tch, tch, why would I be? Rather, let me congratulate you. Your article on William Smith in one of the journals I read recently is so frank, I am surprised.' Fitton's voice is extremely calm.

'I should have written it much earlier, Dr Fitton.'

'I would agree. But he was never found, that's a pity. It's been twelve years now, who knows whether he is dead or alive.'

'Maybe he is alive. We haven't tried our best to locate him, have we?' remarks Farey.

The two men enter the hall. Someone escorts Fitton to a special invitee seat. There are a few more seats vacant. Fitton forces Farey to join him.

The speaker is not known to the two men. They have missed his name because of the delay. He is a young man, and yet extremely eloquent. Fitton looks around to spot a familiar face. Farey presses his arm lightly, 'Listen to this young man, carefully. Seems about to make a revelation.'

Fitton focuses his attention back to the podium.

'…will England ever have a class that will not judge men by their lineage and descent? Who will look at men's work and not at their name and fame? When will the world have a society which will not go only by lineage or aristocratic leanings or even a well-known surname and instead learn more deeply about the person himself? When will we have publishers who will truly know the difference between an original and fake in maps?'

Fitton is curious, 'Who is this?'

Farey whispers in turn to the person on his other side. Then he says to Fitton, 'Apparently a new Professor at Oxford. Buckland has enlisted him in the Geology Department.'

The speaker continues. 'As I have seen in my thirty-year life, this world is infested with jealousy. All around us are busy men, jealous and propelled by ideas of revenge, men who specialise in harassing those far more accomplished than them. So afraid are they of losing power and so filled with insecurity, that they do not wish to give place to any others. My hope lies in appealing to the good sense of a handful of people here. Please tear down the webs that such power-hungry men have built around themselves. Lineage, aristocracy, wealth should not be a standard of measuring good work in the world of science. Let today the Wollaston Medal Committee award ceremony be the harbinger of such novel ideas and change.'

The young man's voice trembles a little towards the end of his speech. He has spoken with emotion today.

Fitton says to Farey, 'Please try to find out his name.'

Farey answers, 'He's a John or Joseph something. The gentleman next to me cannot recall. The boy is new in the circles, so…'

'Ten years ago, one couldn't have believed it possible to deliver a speech such as this. At least we weren't brave enough to do so,' Fitton muses.

The young man concludes his speech. Professor Buckland, University Dean, comes up to the stage and puts his arm gently around the young man. The boy very respectfully moves aside, leaving the dais free for the old Professor.

Buckland begins, 'Thank you, John. Your youth belies your words, and without a doubt, we must all think anew about what you propose. Dr Wollaston had presented several researched facts in a variety of branches of science. But in his death, he asked for a medal to be instituted in the field of geology. Like me, many others were taken aback at his singular request. When I began thinking deeply about this, a strange idea began shaping up within me. I believe strongly that Dr Wollaston was a man with a bent for the natural sciences. His will was made specifically to help bring geology, this new branch of science, to be established as a respected branch within the larger arena. I pay my respect to his farsightedness and to his spirit of fairness.

'What was the study of geology like some twenty-five years ago? It was considered nothing more than eye-pleasing rocks and fossils to beautify our homes. We were living in the assurance of the Biblical claim that the world was created in six days. We were not concerned with what lay under the earth on which we tread so carelessly, why it lay there or even how it came to be? Geology was simply a part of the natural sciences. Just as science had been in the past, a part of philosophy.

'It is because of the relentless effort made by a few that all of us today see revealed the process by which the earth was created. We now know that before man arrived many creatures had walked this very earth. This took place millions and millions of years ago, not only in the last thousand years. They came and lived upon this land, one after the other, lived and birthed. And for some unknown reason, they were wiped off the face of the earth. We do not know yet about the inter-relationships, nor of how they connect with the emergence of man. But I hope that this, too, we will know soon enough. We know that we are the children of Adam and Eve. Whether this is true or false will be determined by our future generations. I may not be fortunate enough to see that day. Death robs us of our bodies. Very few creatures are preserved after death in natural ways. Nature preserves their remains in its own shelves and racks, what we call the sedimentary layers. There is no question of fossils being found in igneous or metamorphic rocks. Almost every sedimentary layer is decorated today with fossils. These fossils and the layers in which they are found are the same age. Like many other formulae, this too is a clue to understand geography and climate. No layer can hold in itself a fossil older than it. Therefore, the layers and the fossils embedded in them must be of the same age. Based on these linkages, a timeline is being created. Scientist Sedgwick and Dr Roderick, who you all know, are here with us today. There is a special person who is associated with them, too. John Phillips. You all just heard him speak. Let me tell you at the outset that due to a lack of funds John could not complete his education uninterrupted. He had to be taken out of school on at least four instances. But let us remember that school does not mean everything. The man to whom John is the deserving successor for his passion to pursue geology is someone I wish to tell you about.

'Let me inform you further that John's schooling followed no formal pattern. He has no mentionable lineage either. He does not belong to a family of dukes, barons, knights, or earls. But his pedigree is to be seen elsewhere, differently. It cannot be determined by accepted social yardsticks. I had not a doubt in inducting John as an Oxford tutor when I found out that his education had been carried on under the aegis of one such who none of us will hesitate at all to pay respect to.

'This is a man who, without any support whatsoever, from either government or private organisations, has toiled alone and continuously for the last sixteen years to bring to us work on the last hundred years. England's scientific community knew of him and yet refused to recognise or honour his work. Because of which the references to certain problematic points in John's speech. Be that as it may, today, we have no difficulty in admitting to this man's greatness, to acknowledging him, and no other, as the father of this new discipline of geology, at least I do not have the strength to deny this.

'I acknowledge, yes, that my knowledge of the field is limited. And yet, from within this limited understanding, I, as well as all the medal committee members have unanimously, and without a single disagreement, chosen this best qualified name. Work can never be recognised with a medal, but that it helps to bring recognition to a name, I do not hesitate to declare. The first medal for research in the geological sciences in keeping with the condition of the will made by the extremely accomplished Dr Wollaston is, therefore, being conferred on a man who should have been introduced to the world long before today, but who we have all failed to honour.

'In this gathering today, with a heart full of joy and relief, I inform you all that the first Wollaston Gold Medal goes to the Father of English Geology, William Smith.'

As soon as the name is announced, Dr Fitton stands up and begins to clap. Farey's expression is inscrutable. He is sitting, bent forward, his head down.

Many members are clapping, although it is not yet a vigorous crescendo. Many among the audience stare at the podium eagerly. Fitton cannot continue to stand anymore. It would be rude. After all, he's in the first row, blocking the view of those behind him. He sits down.

The medal is now brought to the stage. Sedgwick and Roderick are standing there, too.

Fitton asks Farey, 'Are you feeling ill?'

Farey's voice is tired, 'No, I am fine, thank you.'

'Look up now, William Smith is about to arrive.'

Farey looks helplessly at him, 'How is that possible?'

Fitton's expression darkens, 'What do you mean? Is he dead? You knew and yet did not tell me? Dear God!'

Meanwhile, Buckland continues at the dais, 'I'd like to inform all our guests that William Smith is not too well. It is after numerous requests that he has agreed to join us today. Professor John Phillips of the Department of Geology, Oxford

University, is his nephew. Smith undertook the responsibility of his nephew's education as long as he could. For the last twenty years, John has walked alongside his uncle and studied nature in the most determined and careful manner. For the last several years, William Smith has been a resident of Scarborough. Last night he arrived at Prof Phillips's house in London. Many of us may not be aware, the outstanding rotunda shaped museum in Scarborough is designed by William Smith himself. Without wasting any more time, may I invite to the stage, William Smith.'

Fitton clasps his mouth with both his hands, 'Dear God, is this William Smith! Is this his condition now?'

An old, infirm man, shuffling one step at a time, walks into view, leaning heavily on Professor John Phillips. He is bent forward, his frame bony with age. There is perhaps a hint of a smile on his face, and yet it is obvious that even this short walk has tired him out.

Dr Roderick places the heavy gold medal tied up with a bright blue ribbon around his neck. He takes the old man's hand in his own. Then he walks him gently to the edge of the stage, facing everyone.

William Smith's feeble voice is heard, 'Good afternoon. I am happy to be selected for the Wollaston Medal. I thank all the committee members. I thank all those who are present here today.

'Man's relationship with stones is extremely intense. I'll conclude with a small reference. In its true sense, the man who was responsible in starting our journey to understand science, was Isaac Newton. Newton was placed upon a bed of Oolite limestone when he was born. I was thrilled when I first heard about this. That thrill I share, today, with all of you. Thank you.'

The hall now resounds with thunderous clapping.

Dr Fitton comments, 'So the Scarborough Museum designer is Smith after all?'

'Yes, that's what I heard. He had been travelling around all the time these last twenty years. Occasionally, he would visit London, either to meet Cary or Joseph Banks,' Farey answers.

'You knew all this all the while. Then why didn't you tell me?' There is censure in Fitton's voice.

'It would not have helped, Dr Fitton. I've waited long for this day, too.'

'But the man is so ill. How did this come about?'

'Walking on uneven terrain day after day, hammering away at stones. All that has affected his nerves tremendously. Work that would take hundred years, he did in sixteen. This is the price he is paying for his devotion,' Farey speaks softly, halting between his words.

'How do you know so much? I don't think Cary or Joseph Banks would have shared so much information with you.' Fitton blurts out and then looks at Farey, a tad embarrassed.

Farey smiles. 'You are right. They do not think well of me. I became curious when I read about the Scarborough Museum in the newspapers. The way the museum had been planned, it would require a huge number of fossils. It would not be possible for an ordinary person to contribute this. Soon after, I read an advertisement about a monthly workshop on geology in a Scarborough journal. Smith was the force behind these workshops.'

'You visited Scarborough!'

'Yes. I stood by the lecture hall for long. I couldn't gather the courage to go inside and meet him. I saw him being carried by two men into the hall on a stretcher. He could no longer walk. What you saw now is much better than how he had been.'

'Come, let's go to Scarborough,' suggests Fitton.

'But go and congratulate him first. And then you must attend the dinner party,' answers Farey.

They catch a glimpse of Smith as they enter the dinner hall. Smith is leaving, a woman walks alongside him.

Farey says, 'Let's go, Dr Fitton. We need to go far.'

'Will you leave without meeting Mr Smith?'

'Not today, Dr Fitton, it's not my time yet. Please go and meet Prof Phillip. I'll wait for you.'

While shaking Phillips's hand Dr Fitton says, 'What has happened is truly the best. I have hoped for something like this and waited with great eagerness for a long time.'

'I have read all your writing on Uncle Smith. You were the first to make an effort towards getting him honour and deserved recognition,' Phillips answers.

'I am hoping to make a trip to Scarborough for a private meeting.'

'You may not need to do so. I'm arranging for him to come to London.'

'Oh, that would be excellent. There should also be a celebration by the Society, I feel.'

Prof Phillips's face becomes serious and he forces a smile to camouflage his expression.

'He is no longer interested in what the Society wants to do or does not. I cannot promise whether he will join if they do plan something.'

'I agree, it is too late. But we must always extend the chance to make amends. Mr Cary has many of Mr Smith's maps. Have you thought at all about getting those published?' asks Fitton.

'There are talks underway. Whatever is to be done should be done within the next year. He is not too well. And then he is worried about Aunt Ann, too.'

There is a huge crowd waiting to meet Prof Phillips. Dr Fitton bids him goodbye and proceeds to the dining table. Farey joins him now.

'Do you wish to meet Smith?' Farey asks suddenly.

'Did he not leave?'

'No, he went into a room. I have checked it out.'

'You managed to follow him within this time? You are truly...' Fitton stops.

'The right man to snoop around? I do agree. I have been made to thief around all this while, have I not, Dr Fitton?'

'Please, please do not speak like that. Come, let's go find him,' says Fitton and puts his arm around Farey gently.

There is a faint smile on Farey's face. They find Smith sitting in a tiny room. The door is ajar. But they can't go in. Because Smith is sitting on a chair. And right next to him, standing, is a woman, weeping away. Smith is holding her hand.

'Who is she? Smith's wife?' Fitton asks in a low voice.

'No. She used to be Smith's secretary. Almost twenty years ago. Mrs Kitten,' says Farey.

'So much emotion for an old employer! Were they in love?'

'No. But she was very affectionate towards him, like a mother. She is six years his senior.'

'She looks so much younger. How quickly our dear Smith has aged.'

The woman is touching a cross to Smith's forehead and whispering a prayer.

'We cannot meet him today, Dr Fitton. Let them be alone now. If this woman had not helped as bravely as she did, we would never have had a single of Smith's maps with us today. Her contribution is invaluable,' Farey whispers and moves away quickly, walking briskly away, out of the building.

'What a surprise! You seem to know it all. Let's leave them then. We will meet him another time. Prof Phillips did mention that Smith might come to London soon, perhaps permanently,' Dr Fitton says as he follows Farey out.

YEAR 1834, SCARBOROUGH

Dr Fitton is visiting Scarborough on a special assignment. William Smith has stayed with his nephew Professor Phillips in London only for a few days. He has no desire to stay in London permanently. Mrs Mary Ann Smith's medical condition is a big reason for this. To live near an unpopulated shore by the sea helps to bring down the occurrences of her mental bouts. Professor Phillips had hoped to admit Ann for treatment permanently at a mental asylum. But Smith has not agreed, for reasons best known to him. He has brought her home, to be near him always.

It takes Dr Fitton three days to complete his research on the fossils at the Scarborough Museum. Smith's maps are framed on the museum walls all along the staircase. Although Sir Johnston is the overall in-charge of the museum, the person who takes care of the daily details of the place, Williamson, is an extremely accommodating man.

He mentions, 'Smith and his family stayed with us for a long period. The family arrived suddenly, one day, in Scarborough seven or eight years ago. The Yorkshire Mayor assigned the task of surveying the drinking water pipe plan to Smith. At the time, Smith's wife was undergoing treatment for her condition at a mental hospital. At first, when the treatment seemed ineffective, the doctors advised that they spend some time close to the seashore and its refreshing breeze. Mrs Smith has benefited tremendously.'

Dr Fitton says, 'Smith must be very happy with his medal.'

Williamson smiles wanly. 'Is it at all possible to get back Smith's fossils from the British Museum? Do please try if you can. They will not last too long if left there.'

'I don't even know if anyone is making any effort to retrieve them at all or not,' answers Fitton.

'How does one medal help, Dr Fitton? Smith's maps have not yet been recognised and honoured for their singular contribution.'

'It takes time, Mr Williamson. The medal is not any less important, it helps for his name to gain momentum.'

'That is indeed true, and yet there is a time for everything. Had this recognition been conferred some twelve years earlier, a talent would not have gone to waste.'

'I noticed a framed handwritten sheet on one of the museum walls. Would that be in Smith's hand?'

'Yes, a page from his diary. A gift from Professor John Phillips. He is also writing a biography of his uncle.'

'Would it be possible to meet Mr Smith once?' Fitton asks.

'He lives in Sheffield. It's at quite a distance from Scarborough. I'll return home today. You could come with me. You could put up as my guest tonight.'

'Wonderful, let's proceed in that case.' Dr Fitton is doing his best to control his excitement and yet it spills over in his voice.

After driving on for close to an hour and fifteen minutes, Williamson brings his buggy to a stop by a large field.

He says, 'We are here. Come down, Doctor. Let the horses rest a while.'

Fitton gets off. He is surprised. 'I do not see your house anywhere here.'

'The one you have come to see is sitting there.' Williamson points to a distance.

Two people are sitting by the corn field. A woman, her body worn out, almost reclines on her chair. And next to her, an old man, his face covered in white beard. He seems close to seventy, even eighty perhaps. He's holding a cup close to the woman's mouth. Maybe it contains tea, or coffee. Or hot water!

The faces are not too clear from this distance.

'Who are they?' Fitton asks.

'William Smith and his wife, Mary Ann.'

'I saw him about two years back, at Oxford. He seems much older now, very sick and infirm. He should not be more than sixty-four or sixty-five years, right?'

'Every morning after breakfast, they are both brought here. They look after the fields all day long. He doesn't want his wife to be left alone at home. The lady too is happy sitting next to her husband. The village people will come and take them back home before it turns dark.'

'Looks after the fields? Such a renowned scientist, to look after the fields? The man who has designed such a spectacular museum for Scarborough? You could have allowed him to supervise the running of the museum instead,' Fitton suggests.

Williamson is calm. 'The museum was ready ten years ago. Time flies, Doctor. It's been a long while since he has managed to walk around comfortably. He would visit a few universities and seminars for talks in the early days. Of course, in the hope to make some money. As soon as John became employed, Smith stopped all that. Our King, the King of England, William the IV, has arranged for a pension for Smith to the tune of one hundred pounds a year. They live comfortably with that amount in the village. These days he doesn't like to talk much. But our farmers have benefited tremendously with his presence. And he is very happy, sharing his knowledge about the quality of soil here, the seed care required, and with making suggestions about where to do mono-cropping or bi-cropping. Anything related to farming.'

'What are you saying! We need this man even now for the cause of science. So many in the business of minerals will be helped, innumerable students will be encouraged to do research. Arrangements must immediately be made to take him back to London.' Dr Fitton's voice has taken on a sharpness.

'Are you suggesting that no one has tried it, not even his nephew? Every one of us has done so, but he is not willing at all. He introduces himself as a farmer whenever someone new comes to the village. He sits frozen when he hears anything to do with geology. We do not disturb him anymore.'

'Should I try to speak with him?'

'Let him live as he wishes to, Doctor. Please do not try to convince him of anything, there is no point. And then it may not at all be pleasant for a member of the Society to speak with him. What if he asks suddenly about what has happened to his maps, or about the condition of his fossils at the British Museum? Will you be able to answer him?' Williamson asks, one question at a time, almost as if firing away.

Dr Fitton sighs long. And then, without a word, he comes to sit in the buggy. The sun is high up, sitting right on top of the corn plant tips. The lines written on that diary sheet framed on the Scarborough Museum walls were:

This map stands now next to all my work, towering like a ghost, laughing, its teeth all bare. I am helpless, I cannot help myself. I could not overcome the fate of my birth.

Perhaps Mary Ann has finished eating her food. The old man wipes her face clean with a scrap of cloth. He seems to do it with a deep compassion. The woman puts her head through her husband's loving arms and tries to steal a glance at the dying sun.[1]

1 William Smith's collection of rocks and fossils are kept safely and with great care in a vault in the British Museum. One requires special permission to have a look at the collection. In the year 1860, the Geological Society of London acknowledged the work done by William Smith. Smith had died twenty-one years before this happened. Before his death, he had come to terms with the fact that his maps would not ever carry his name as their creator.

ACKNOWLEDGEMENTS

This book has meant the world for several deeply personal reasons. I am fortunate it made its way to me, and for this, it will always have my gratitude. To many others as well, my deepest thanks.

To William Smith, for his life of amazing grit and perseverance, and his reassurance from across the years that gives fresh meaning to passion and patience.

To Maitry Roy Moulik, for bringing William Smith alive with such ease and sensitivity, and for making an old story new.

To Nirmal Kanti Bhattacharjee, for thinking of me to tell this powerful story all over again.

To Niyogi Books and its fabulous team, for being there, always. To Antra K., for the book's dramatic cover.

And finally, to Arvind Krishna Mehrotra, for trusting us with his poem "Ramapithecus and I," so startlingly apt as an Epigraph for this book.

Chandana Dutta